# TABLE OF CONTENTS

| | |
|---|---:|
| **THE INVISIBLE GOVERNMENT** | 1 |
| • The U.S. Government Counter Intelligence Organizational Chart | 2 |
| • U.S. Space Counter Intelligence HQ, Organizational Chart | 2 |
| • Operation Majority | 4 |
| • The U.S.'s Alien Cover-up | 5 |
| • Top Secret - Majic | 8 |
| • The "Majic Projects" | 9 |
| • Majic's Contingency Plans | 10 |
| • Majic's Secret Weapons Against the Aliens | 11 |
| • Information from Inside Report Number 13 | 12 |
|    • Other Majic - Alien Tips from inside Report Number 13 | 12 |
| • NSA Security System | 14 |
| | |
| **FACTS ABOUT ALIENS ON EARTH (1) - CODE: OMNIDATA** | 15 |
| • Facts About Aliens on Earth (1) | 15 |
| • Main Important UFO Crashes, Under Ultra-Secret Cover-ups | 16 |
| • Extra-Terrestrial Classifications | 16 |
| • Alien Presence on Earth | 17 |
| | |
| **CATTLE MUTILATIONS** | 18 |
| • My General Personal Observations | 18 |
|    • Human Reasons for the Mutilations | 19 |
|    • Alien Reasons for the Mutilations | 19 |
| • Black Helecopter Involvement in Mutilations | 19 |
| | |
| **THE ANTARIAN CONNECTION - CODE: OMNICORD** | 21 |
| • Time and UFOs | 21 |
| | |
| **CROP CIRCLES PHENOMENA - CODE: SEC** | 23 |
| • Crop Phenomena: Circles Affair | 23 |

## THE DULCE BASE - CODE: J.B. III    26
- The Dulce Base    26
- Caught in the Game    26
- Who's Planet is this ?    27
- The Secret Activity    27
- R & D and the Military Industrial Complex    27
- Surviving the Future    28
- The Transamerican Underground Subshuttle System, (T.A.U.S.S.)    28
- Overt and Covert Research    29
- Each Base has its own Symbol    31
- Inside the Dulce Base    31
- The Town of Dulce    32
- Mind Manipulation Experiments    33
- Better Living Through Bio-Tech ?    34

## SOME KINDS OF ALIEN LIFE FORMS WE KNOW ABOUT, (An Anthropological Analysis)    36
- U.S. Government Data from Autopsies of Aliens    36
  - Uni-Terrestrial Data, Code: UNA    37
    - Mission Skills and Assignments Chart    40
    - Mission    40
  - Ultra-Terrestrial Data, Code: ULTRON    42
  - Intra-Terrestrial Data, Code: INTERAV    44
  - Meta-Terrestrial Data, Code: MEREDITH    54
  - Para-Terrestrail Data, Code: PARAMUS    55
  - Greys    58
    - Extra-Terrestrail Data, Code: EBEs    58
      - The First EBE    58
      - KRLL    58
    - Rigelians    60
      - The Rigelian Saga    62
      - Alien Bases at Sol III Perimeter    63

- Project BETA, The Study of Grey Psychology ... 66
- Taxonomy of some Extra-Terrestrial Humanoids ... 67
- Greys Data from Autopsies ... 68
  - Type A ... 68
  - Type B ... 70
  - Type C ... 72
- Nordics Data from Autopsies
  - Type D ... 74
- Oranges Data from Autopsies
  - Type E ... 75

## ALIEN TECHNOLOGY ... 76
- Alien Crafts with Dimensional Factors and/or Dimensional Origin ... 76
- The U.S. Government and Alien Technology ... 76
- The Implant or S.B.M.C.D. ... 77
- Aliens and the Electromagnetic Spectrum ... 78
- Alien Artifacts used at Mutilatons ... 79
  - The Biological Scanner ... 79
  - The Spray Applicator ... 80
  - The Field Reader Tube ... 81
  - Surgical Scalpels ... 81
  - Skin-Grafting Lasers ... 81
  - Blood Attracting Device ... 81

## U.S. GOVERNMENT SECRET TECHNOLOGY ... 82
- Mystery Helicopters and Mutilations ... 82
- Surveillance Devices ... 83
  - The RPV, AROD ... 83
  - The Hale, ADR/238F ... 84
  - The TRA, SN-75, XH-75D ... 84

## ALIEN ABDUCTIONS ... 86
- "The Monitors" ... 86
- The Case for Alien Abductions ... 86

- The Most Common Areas Examined by the Aliens  90
- Biological Specimens, Samples usually taken for Abductees and/or Witnesses  93
- Pregnancy  94
- Our Genetic Code is under Siege  96
- The Metagene Factor  98

## INCULCATION DEVICES AND
## THE PRACTICE OF MANIPULATION OF THE MIND - CODE: NAR  104
- Learning, Teaching and Assimilation techniques of electronic space societies  104
- Inculcation Devices  105
  - Inculcation Bar Device  105
  - Inculcation Monitor  105
  - Catecholine Beta-Lipotropin 9753  105
  - Rapid Inculcation Processes  105
  - Ship Diagrams

# CODE: NSC

AC and the Government
23-C

## U.S. COUNTER INTELLIGENCE ORGANIZATIONAL CHART

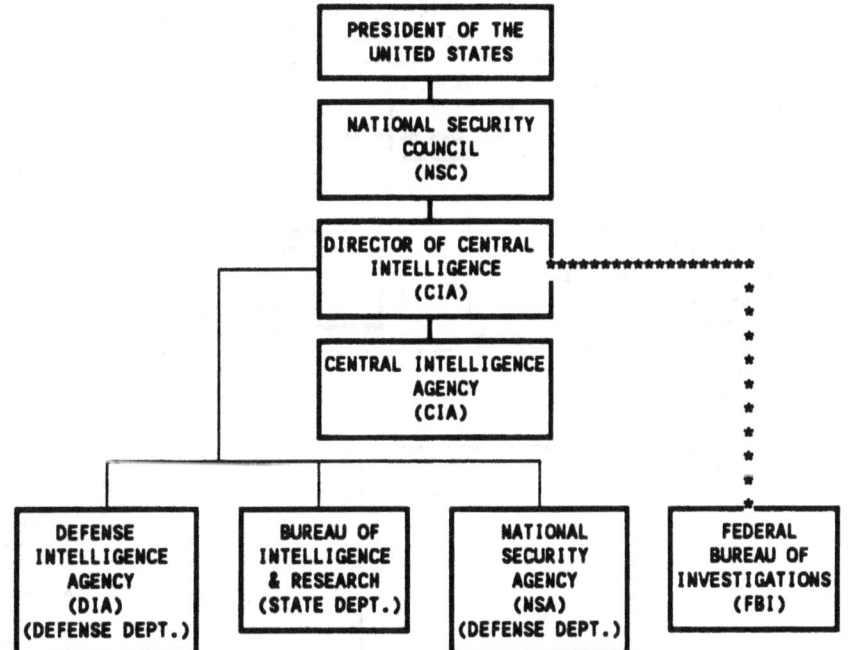

———— Control

———— Coordination; Control of Budgetary Resources

****** Coordination Only

# STAR WARS CITY (SDI/01)

**Code:** Strategic Defense Initiative (SDI) also known as Spaceship Defense Initiative (Star Wars Project).
**Control Location:** Cheyenne Mountain, Colorado Springs.

## U.S. SPACE COUNTER INTELLIGENCE HEAD QUARTERS

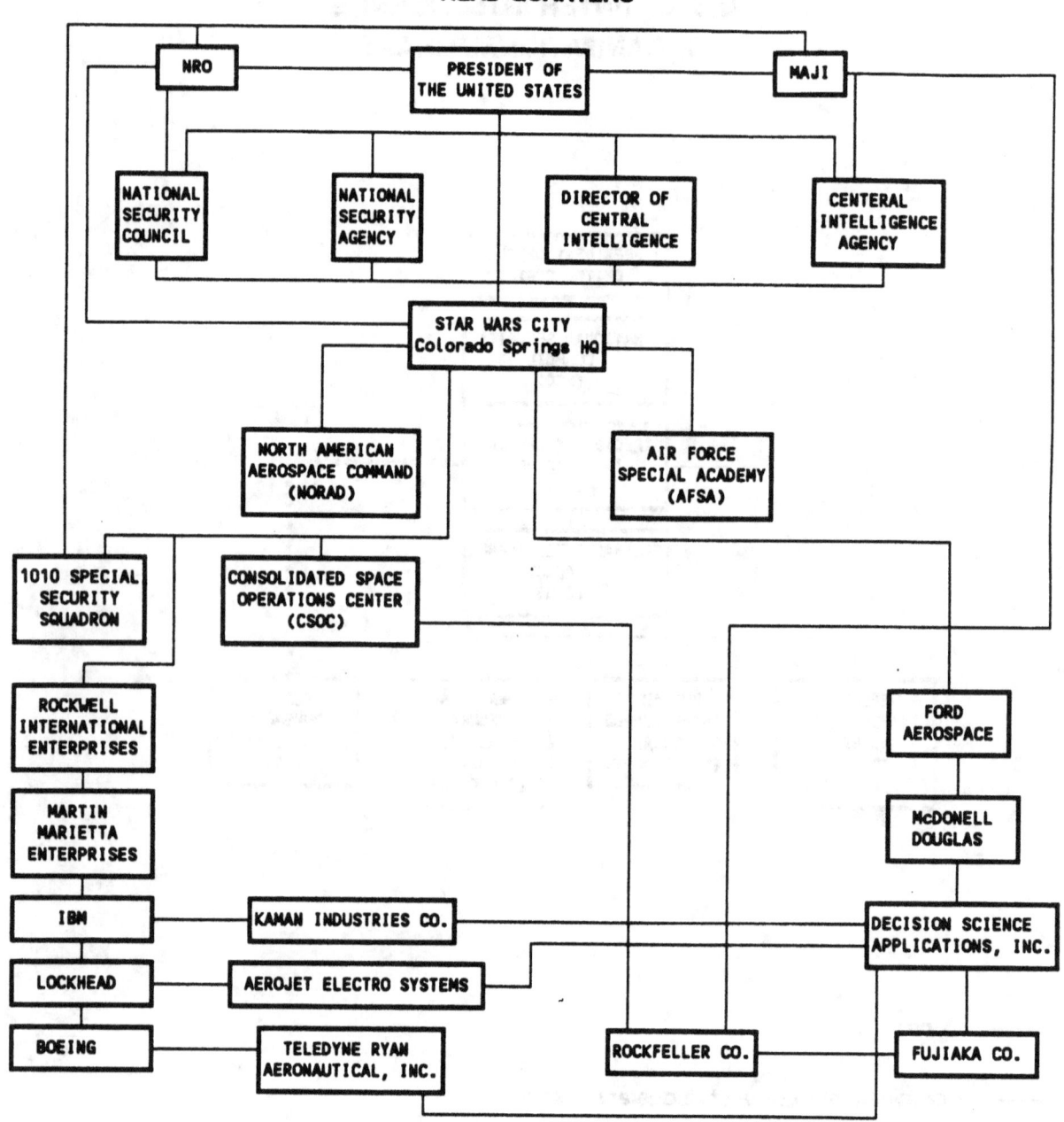

ADR/23 - NSC OCM A7BMC - CODE: MIRAMA 55A - ACL - 3 65ATN-6  (MOZART) # 6.2

## THE INVISIBLE GOVERNMENT

The National Security Council, Old Executive Office Building, Washington, DC. 20506, Phone Number (202) 395-4974, was established by the National Security ACT of 1947 (61 Stat. 496;50 U.S.C. 402). President Truman signed the National Security ACT on July 26, 1947 and immediately named Secretary of the Navy Forrestal as the first Secretary of Defense. Forrestal and others were sworn in September 17, 1947.

The ACT established "under the National Security Council" a Central Intelligence Agency headed at that time by "Director", Rear Admiral Roscoe Hillenkoeter. It provided a comprehensive program for future Security of the United States, and the ACT created the NSC to advise the President with respect to the integration of domestic, foreign, and military policies relating to the National Security, with the special duty to "Access and Appraise the objectives, commitments, and Risks.... The funds for the CIA were hidden in the annual appropriations for other agencies. Today, collectively, U.S. Intelligence Operations (almost totally clothed in secrecy) cost more than $1 Billion Dollars annually.

The researchers Moore, Shandera, and Friedman have referred to briefing document:

*Operation Majestic 12*
*NSC, MJ-12*
*Special Studies Project*

MJ-12 is said to be a TOP SECRET Research and Development, Intelligence Operation established by President Truman on September 24, 1947. MJ-12 was a "Committee" set up inside the NSC. In 1954, President Eisenhower signed the Secret Executive Order, "Order Number 54-12".

> *Newsweek of June 22, 1964, carried a review of "The Invisible Government"......"One of their major revelations is the existence of the Special Group "54-12", a hitherto classified adjunct of the National Security Council, specially changed by the President with ruling on special operations."*

The NSC called this group the "54-12" committee "which gave the President responsibility of approving all "Black" covert projects. This committee has undergone several changes over the years, and since then, has been called the "SPECIAL COMMITTEE", the "303 COMMITTEE and currently the "40 COMMITTEE" and is represented as XXXX = the double - double cross. It is described as the "Directorate" of the NSC. The "40 COMMITTEE" has access to advanced technology and teams to cover-up, "the cover-ups"! In the past, this committee was headed by Dr. Henry Kissinger (code name: "The Overseer") when he gave William Colby permission to commission Howard Hughes, Summa Corporation to build a submarine, and a special craft or salvage vessel. The salvage vessel called "Glomar Explorer", was equipped with refrigeration capacity for up to a hundred bodies. (Why???)

In early May 1988, the former President Ronald Reagan said he often wonders what would happen if the Earth was invaded by a "Power from Outer Space". It has also been reported that the President is briefed on UFO developments and "Alien Visitors, by Planetary Intelligence group #40 (PI-40).

**OPERATION MAJORITY - CR-20M7/6.2 - FILE: MTR/K-25 MWC/JL - AFMWC/1972 USN**
Operation Majority is the name of the operation responsible for every aspect, project and all consequence of Alien presence on Earth.

**Majesty** was listed as the code word for the President of the United States for communications concerning this information.

**Grudge** contains 16 volumes of documented information collected from the beginning of the United States investigation of the Unidentified Flying Objects (UFOs) and Identified Alien Crafts (IACs). The project was funded by the CIA, (confidential funds, non-appropriated) and money from the illicit drug trade. Participation in the illegal drug trade was justified in that it would identify and eliminate the weak elements of our society.

The purpose of project Grudge was to collect all scientific, technological, medical and intelligence information from UFO & IAC sightings as well as contacts with Alien Life Forms. This orderly file of collected information has been used to advance the United States Air Force Space Program, (TOP SECRET).

**Jason Society (Jason Scholars)** President Eisenhower (former) commissioned a secret society known as the Jason Society (or Jason Scholars) under the leadership of the following; Director of Central Intelligence, Allen Welsh Dulles, Dr. Zbigniew Brzezinski, President of the Trilateral Commission from 1973 until 1976, and Dr. Henry Kissinger, leader of the scientific effort, to sift through all the facts, evidence, technology, lies and deceptions and find the truth of the Alien question. The society was made up of thirty two (32) of the most prominent men in the U.S.A.

**MJ-12** is the name of the secret control group inside the Jason Society. The top twelve (12) members of the thirty two (32) members of the Jason Society were designated the MJ-12. MJ-12 has total control of everything. They are designated by the code J-1, J-2, J-3, etc. all the way through the members of the Jason Society. The director of Central Intelligence was appointed J-1 and is the Director of the MJ-12 group. MJ-12 is responsible only to the President of the United States. Believe it or not, MJ-12 runs most of the worlds illegal drug trade. This was done to hide funding and thus keep the secret from the Congress and the people of the United States. It was justified in that it would identify and eliminate the weak and undesired elements of our society.

\* The actual cost of funding the Alien connected projects is higher than anything you can imagine !

It was the MJ-12 group who ordered the assassination of President Kennedy when he informed MJ-12 that he was going to tell the public all the facts of the Alien presence. He was killed (coup de grace) by the Secret Service Agent driving his car, (who fired the final and most critical shot) and it is plainly visible in the film, which has been held from public view, and the twenty two (22) material witnesses to the Kennedy's murder died within two (2) years thereafter.

MJ-12 has a special program for elimination of the considered weak members from U.S.A's society.

       **ON A BIG SCALE**....Creation of an artificial disease known as "**AID's**".
       **ACCIDENTS**....of people who we know to have been secretly eliminated.
**DISAPPEARANCES/MURDER**....of some people who we know to have been secret eliminations also.

The above, are just a few of the tactics which MJ-12 uses when they consider people Dangerous by knowing too much !

A secret meeting place was constructed for the MJ-12 group in Maryland and was described as only accessible by air. It contains full living, recreational, and other facilities for the MJ-12 group and the Jason Society. It is code named "The Country Club". The land for The Country Club was donated by the Rockefeller Family. Only those with TOP SECRET - MAJIC clearances are allowed to go there.

**MAJI** = Majority Agency for Joint Intelligence. All information, disinformation, and intelligence is gathered and evaluated by this Agency. This agency is responsible for all disinformation and operates in conjunction with the CIA, NSA, DIA, and The Office of Naval Intelligence. This is a very powerful organization and all Alien projects are under its control. MAJI is responsible only to MJ-12. MAJIC is the security classification and clearance of all Alien connected material, projects, and information.

   **MAJIC** = means MAJI controlled, (MAJI plus Controlled = MAJIC).

**NOTES:** MJ-1 is the classification for the director of MAJI, who is the Director of the CIA and reports only to the President. Other members of MAJI are designated MJ-2, MJ-3, MJ-4, etc. This is why there is some confusion about references of MJ-12, the group or MJ-12 the person.

- Designation for MJ-12, the group are MAJI or MAJIC.
- Designation in official documents about MJ-12 means the person only.

**THE U.S.'s ALIEN COVER-UP** - It seemingly all began thousands of years ago, but for the purpose of this discussion, let's start with some events that we all are familiar with. In 1947, two years after we set off the first nuclear explosion that our current civilization detonated, came the Cap-Mantell episode, where we had the first incident of a military confrontation with extra-terrestrial that resulted in the death of the military pilot. It is quite evident now that our Government did not know quite how to handle the situation. In 1952, the nation's capital was overflown by a series of disks. It was this event which led to the involvement of the United States Security Forces (CIA, DIA, NSA, & FBI) to try to keep the situation under control until they could understand what was happening. During this period, the original members of the group were:

> Secretary James Forrestal
> Admiral Roscoe H. Hillenkoetter
> General Nathan P. Twining
> General Hoyt S. Vandenburg
> General Robert M. Montaque
> Dr. Vanevar Bush
> Dr. Detlev Bronk
> Dr. Jerome Hunsaker
> Dr. Donald Menzel
> Dr. Lloyd V. Berkner
> Mr. Sidney W. Souers
> Mr. Gordon Gray

This group was a working group, established by the Government, known as MJ-12.

The MJ-12 group has been an existing group since it was created with new members replacing others that died. For example, when Secretary Forrestal was upset at seeing the United States sold out in World War II, he wound up being sent to a Naval hospital for emotional strain. Before relatives could get to him, he "Jumped Out a 16th Story Window". People close to him consider his suicide contrived. When Forestal died, he was replaced by General Walter B. Smith.

In December of 1947, PROJECT SIGN was created to acquire as much information as possible about UFOs, their performance characteristics and their purposes. In order to preserve security, liaison between Project Sign and MJ-12 was limited to two (2) individuals within the intelligence division of the Air Material Command whose role was to pass along certain types of information through channels. Project Sign evolved PROJECT GRUDGE in December 1948. Project Grudge had an overt civilian counterpart named PROJECT BLUE BOOK, with which we are all familiar. Only "Safe" reports were passed to Blue Book.

MJ-12 was originally organized by General George C. Marshall in July 1947 to study the Roswell-Magdalena UFO crash recovery and debris. Admiral Hillenkoetter, director of the CIA from May 1, 1947 until September 1950, decided to activate the "ROBERTSON PANEL", which was designed to monitor civilian UFO study groups that were appearing all over the country. He also joined NICAP in 1956 and was chosen as a member of its board of directors. It was from this position that he was able to act as the MJ-12 "Mole", along with his team of other covert experts. They were able to steer NICAP in any direction they wanted to go. With the "Flying Saucer Program" under complete control of MJ-12 and with the physical evidence hidden away, General Marshall felt more at ease with this very bizarre situation. These men and their successors have most successfully kept most of the public fooled for at least 39 years or more, including much of the western world, by setting up false experts as real experts and throwing their influence behind them to make their plan work, with considerable success, until now.

Within six (6) months of the Roswell crash on July 2, 1947 and the finding of another crashed UFO at San Augustine Flats near Magdelena, New Mexico on July 3, 1947, a great deal of reorganization of agencies and shuffling of people took place. The main thrust behind the original "Security Lid", and the very reason for its construction, was the analysis and attempted duplication of the technologies of the discs. The activity is headed up by the following groups:

> *The Research and Development Board (R&DB)*
> *Air Force Research and Development (AFRD)*
> *The Office of Naval Research (ONR)*
> *CIA Office of Scientific Intelligence (CIA-OSI)*
> *NSA Office of Scientific Intelligence (NSA-OSI)*

No single one of these groups were supposed to know the whole story. Each group was to know only the parts that MJ-12 allowed them to know. MJ-12 also operates through the various civilian intelligence and investigative groups. The CIA and the FBI are manipulated by MJ-12 to carry out their purposes. The NSA was created in the first place to protect the secret of the recovered flying discs, and eventually got complete control over all communication intelligence. This control allows the NSA to monitor any individual through mail, telephone, telexes, telegrams, and now through on-line computers, monitoring private and personal communications as they may desire.

TOP SECRET - MAJIC
SUBJECT: MAJI
PROJECT: GRUDGE/AQUARIUS : (TS/MAJIC)
   Document Control: ECN
   Classified By : MJ1/MAJI
   Declassified on : Exempt

**PROJECT AQUARIUS**
(TS/ORCON), (Proword): Grudge contains sixteen (16) volumes of documented information collected from the beginning of the United States Investigation on Unidentified Flying Objects, (UFOs) and Identified Alien Crafts, (IACs). The project was originally established in 1953, by order of President Eisenhower, under control of the CIA and MAJI. In 1960 the project's name was changed from project SIGN to project AQUARIUS. The project was funded by CIA confidential funds (non-appropriated). The project assumed full responsibility for investigation and intelligence of UFOs and/or IACs, after December 1969 when Project Grudge and Blue Book were closed. The purpose of project Aquarius was to collect all scientific, technological, medical and intelligence information from UFO and IAC sightings and contacts with Alien Life Forms. These orderly files of collected information have been used to advance the United States Air Force Space Program (not NASA).

Aquarius is a project which compiled the history of Alien presence and their interaction with HOMO SAPIENS upon this planet for the last 25,000 years and culminating with the BASQUE PEOPLE (PAIS BASCO) who live in the mountainous country on the border of France and Spain and the Assyrians (or Syrians, originally from the Syrius Star).

(TS/ORCOM) The preceding briefing is a historical account of the United States Government's Investigation of Aerial Phenomena, Recovered Alien Aircraft, and contacts with Extra-Terrestrial Life Forms.

**THE PROJECTS UNDER "PROJECT AQUARIUS"**

1. (TS/ORCOM) PROJECT PLATO: (Proword: Aquarius) Originally established as part of Project SIGN in 1954, its mission was to establish Diplomatic Relations with Aliens. This project was successful when mutually acceptable terms were agreed upon. These terms involved the exchange of technology for secrecy of Alien presence and non interference in Alien affairs. Aliens agreed to provide MAJI with a list of Human contacts on periodic basis. This project is continuing at a site in New Mexico.

2. (TS/ORCOM) PROJECT SIGMA: (Proword: Aquarius) Originally established as part of Project SIGN in 1954. Became a separate project in 1976. Its mission was to establish communication with the Aliens. This project met with positive success (SIC). In 1959, the United States established primitive communications with Aliens. On April 25, 1964 a USAF Intelligence Officer met with Aliens at Mollomar Air Force Base, New Mexico. The contact lasted for approximately

three hours, after several attempted methods of communications the Intelligence Officer managed to exchange basic information with the Aliens. This project is continuing at a site in New Mexico.

3. **(TS/ORCOM) PROJECT REDLIGHT:** (Proword: Grudge) Originally established in 1954. Its mission was to test and fly a recovered Alien Aircraft. First attempts resulted in the destruction of the craft and the death of the pilot. This project was resumed in 1972. This project is continuing in Nevada.

4. **(TS/ORCOM) PROJECT SNOWBIRD:** (Proword: Redlight) Originally established in 1954. Its mission was to develop, using conventional technology, and fly a "Flying Saucer" type craft for the public. This project was successful when a craft was built and flown in front of the PRESS. This project was used to explain UFO sightings and to divert public's attention from project Redlight.

**THE "MAJIC PROJECTS"**

1. **SIGMA** is the project which first established communications with the Aliens and is still responsible for communications.

2. **PLATO** is the project responsible for Diplomatic Relations with the Aliens. This project secured a formal treaty (illegal under the U.S. Constitution) with the Aliens.

> *The terms were that the Aliens would give us "our Government" technology and would not interfere in our History. In return we "our Government" agreed to keep their presence on Earth a secret, not to interfere in any way with their actions, and to allow them to abduct humans and animals.*

The Aliens agreed to furnish MJ-12 with a list of abductees on a periodic basis for Governmental control of their experiments with the abductees.

3. **AQUARIUS** is the project which compiled the history of the Alien presence and interaction on Earth and the HOMO SAPIENS.

4. **GARNET** is the project responsible for control of all information and documents regarding the Alien subjects and accountability of their information and documents.

5. **PLUTO** is a project responsible for evaluating all UFO and IAC information pertaining to Space technology.

6. **POUNCE** project was formed to recover all downed and/or crashed craft and Aliens. This project provided cover stories and operations to mask the true endeavor, whenever necessary. Covers which have been used were crashed experimental Aircraft, Construction, Mining, etc. This project has been successful and is ongoing today.

7. **NRO** is the National Recon Organization based at Fort Carson, Colorado. It's responsible for security on all Alien or Alien Spacecraft connected to the projects.

8. **DELTA** is the designation for the specific arm of NRO which is especially trained and tasked with security of all MAJIC projects. It's a security team and task force from NRO especially trained to provide Alien tasked projects and LUNA security (also has the CODE NAME: "MEN IN BLACK"). This project is still ongoing.

9. **BLUE TEAM** is the first project responsible for reaction and/or recovery of downed and/or crashed Alien craft and/or Aliens. This was a U.S. Air Force Material Command project.

10. **SIGN** is the second project responsible for collection of Intelligence and determining whether Alien presence constituted a threat to the U.S. National Security. SIGN absorbed the BLUE TEAM project. This was a U.S. Air Force and CIA project.

11. **REDLIGHT** was the project to test fly recovered Alien craft. This project was postponed after every attempt resulted in the destruction of the craft and death of the pilots. This project was carried out at AREA 51, Groom Lake, (Dreamland) in Nevada.

    Project Redlight was resumed in 1972. This project has ben partially successful. UFO sightings of craft accompanied by Black Helicopters are project Redlight assets. This project in now ongoing at Area 51 in Nevada.

12. **SNOWBIRD** was established as a cover for project Redlight. A "Flying Saucer" type craft was built using conventional technology. It was unveiled to the PRESS and flown in public on several occasions. The purpose was to explain accidental sightings or disclosures of Redlight as having been the Snowbird crafts. This was a very successful disinformation operation. This project is only activated when needed. This deception has not been used for many years. This project is currently in mothballs, until it is needed again.

13. **BLUE BOOK** was a U.S. Air Force, UFO, and Alien Intelligence collection and disinformation project. This project was terminated and its collected information and duties were absorbed by project Aquarius. A classified report named "Grudge/Blue Book, Report Number 13" is the only significant information derived from the project and is unavailable to the public, (from what I read before from other sources, this Report Number 13, talked about everything inside the Grudge history).

**MAJIC's CONTINGENCY PLANS** - In 1949, MJ-12 evolved an initial plan of contingency called MJ-1949-04P/78 that was to make allowance for public disclosure of some data should the necessity present itself.

- General Doolittle made a prediction that one day we would have to reckon with the Aliens and the Grudge/Blue Book, Report Number 13 stated that it appeared that General Doolittle was correct.

- The Grudge/Blue Book Report Number 13 also mentioned that the document stated that many military government personnel and civilians have been Terminated, (murdered without Due Process of Law) when they had attempted to reveal the secret.

**PLAN A CONTAIN OR DELAY RELEASE OF INFORMATION - CODE: MAJIC/STR/M    03CCPN24**
This plan called for the use of MAJESTIC TWELVE as a disinformation ploy to delay and confuse the release of information should anyone get close to the truth. It was selected because the similarity of spelling and the similarity to MJ-12. It was designed to confuse memory and to result in a fruitless search for material which did not exist. (i.e. William Moore's Documents ?)

**PLAN B SHOULD THE INFORMATION BECOME PUBLIC OR SHOULD THE ALIENS TAKE OVER.**
This plan called for a public announcement that a terrorist group had entered the United States with an Atomic weapon. It would be announced that the terrorists planned to detonate the weapon in a major city. Martial Law would be declared and all persons with implants would be activated by the Aliens. That person, in this specific case, would be rounded up by MAJIC along with all dissidents and would be placed into concentration camps. The PRESS, the Radio and TV would be nationalized and controlled. Anyone attempting to resist would be arrested or killed.

**MAJIC's SECRET WEAPONS AGAINST THE ALIENS** - Although better weapons are being developed every day, the following are a few of the originals created.
**GABRIEL** is a project to develop a High Frequency pulsed sound generating weapon. It was said that this weapon would be effective against the Alien crafts and their Beam weapons. Specifications about the GABRIEL project shows high frequency microwave projections included.

**JOSHUA** is a project to develop a Low Frequency pulsed sound generating weapon. The weapon was developed and assembled at Ling Tempco Vought in Anaheim, California. It was described as being able to totally level any man made structure from a distance of two (2) miles. It was tested at White Sands Proving Grounds. It was developed between 1975 and 1978. It is a long horn shaped device connected to a computer and amplifiers. Ling Tempco Vought (LTV) has since moved from its Anaheim Facility which was near the Grand Hotel across the street from Disneyland. Specifications about the JOSHUA project show low frequency generated with microwave projection included.

**EXCALIBUR** is a weapon to destroy the alien underground bases. It is to be a missile capable of penetrating 1,000 meters of Tufa/Hard packed soil, such as that found in New Mexico with no operational damage. Missile apogee not to exceed 30,000 feet Above Ground Level (AGL) and impact must not deviate in excess of 50 meters from designated target. The device will carry a one (1) to ten (10) Megaton Nuclear Warhead. The secret for a self contained missile "drill" a vertical shaft over 1,000 meters deep consists in a energosintetizer macrowave deflector in the "Missile Warhead".

**OBS** - MAJIC has five (5) other major weapons to use against Aliens but I don't have the data on or about these weapons yet.

## INFORMATION FROM INSIDE GRUDGE/BLUE BOOK, REPORT NUMBER 13

**EBE** is the name or designation given to the live Alien captured at the 1947 Roswell, New Mexico crash. He died in captivity. EBE means Extra-Terrestrial Biological Entity.

**KRLL** was the first Alien Ambassador to the United States of America.

**GUESTS** were Aliens exchanged for Humans who gave us the balance of the YELLOW BOOK. At the onset in 1972 there were only three (3) left alive, now we have around 4,000. They were called Alien Life Forms (ALFs) or OBS. They have a big tendency to LIE.

**YELLOW BOOK** is all that we know about Alien technology, culture and their history.

**RELIGION** - The Aliens believe in a Universal Cosmic God. The Aliens claim that MEN are Hybrids who were created by them. They claim all religion was created by them to hasten the formation of a Civilized Culture and to control the Human Race. They claim that JESUS was a product of their efforts. The Aliens have furnished proof of their claims and have a "device" that allows them to show audibly and visually any part of History that they or we wish to see.

**ALIEN BASES** exist in the four corner areas of Utah, Colorado, New Mexico and Arizona. Six bases were described in 1972, all on Indian Reservations, and all in the four corner area. The base near Dulce was one of them, (also bases at California, Nevada, Texas, Florida, Maine, Georgia and Alaska.

**CRAFT RECOVERIES** - The documents stated that many craft had been recovered. The early ones from Roswell, Aztec, Roswell again, Texas, Mexico and other places, which will be discussed later.

**ABDUCTIONS** were occurring long before 1972, early civilizations refer to these incidents. The document stated that humans and animals were being abducted and or mutilated. Many vanished without a trace. They were taking sperm and OVA samples, tissues, performed surgical operations, implanted a spherical device, (40 to 50 microns in size near the optic nerve in the brain) and all attempts to remove the device have resulted in the death of the patient. The document estimated that one (1) in every 40 people had been implanted. This implant was said to give the Aliens total control of the Human.

## OTHER MAJIC - ALIENS TIPS FROM REPORT 13 - SRD/28 CODE: ABRAMA 33A/C

- We confirmed in 1989, the existence of Alien crafts at a hangar on Edwards Air Force Base. The hangar is at North end of the base. It has been guarded by non-Edward's personnel, who are NRO-DELTA personnel. The guards wear a badge that is red with a black triangle on the face of the badge. No one was allowed near the hangar without this badge.

The NRO-DELTA personnel are no longer guarding the hangar, Edwards security forces are are guarding it now, and they are instructed to check the hangar each hour and report the status NRO. In addition, they have been instructed never to enter the hangar, even if it has been broken into. The hangar is still locked and no one is allowed inside without special authorization from NRO/DELTA.

We also have confirmed the existence of Alien Materials at another Special hangar at Edward's AFB.

- The badge insignias that are on some Alien crafts and Alien Flags, is called a TRILATERAL INSIGNIA (TRIADE).

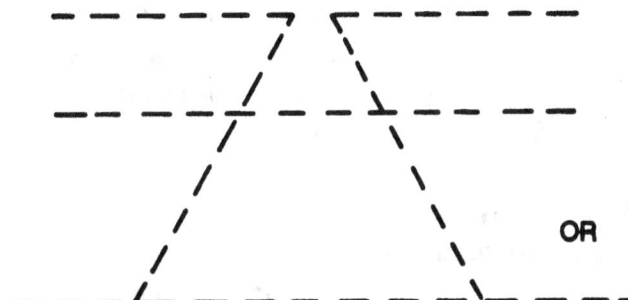

OR

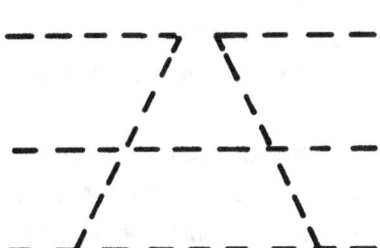

Found on some Spacecrafts

Marks bases and Landing sites.
This symbol is visible,
"Only when viewed from Directly Overhead"

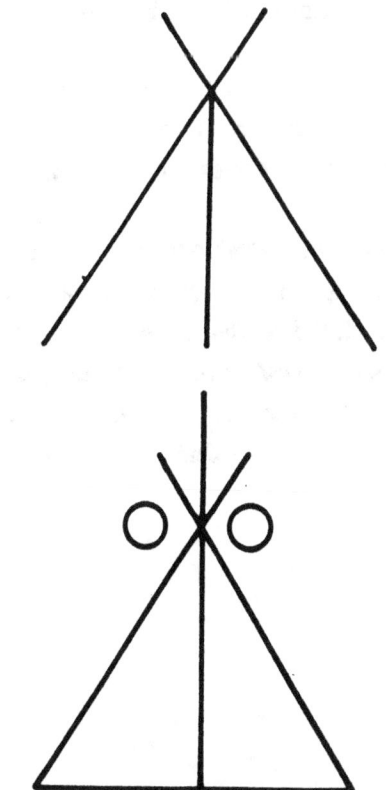

This symbol is found on some Regelians
Spacecraft and their Uniforms

- **LUNA-2** - Code name for the location of the second most important underground Rigelian base in New Mexico. The base is Alien controlled and NRO, DELTA, and Alien protected. Luna is ongoing.

- **FAR SIDE OF THE MOON** - Is the term used in reference to inside LUNA-2. The LUNA-2 underground base, UFOlogists and some other various people who have stumbled upon this term have a misunderstanding and believed that it was a reference to LUNA-1, the moon base.

**NSA Security System** - The National Security Agency, (NSA) was created to protect the secretly recovered flying disks, and eventually got compete control over all Communication Intelligence. This control allows the NSA to monitor any individual through the mail, telephone, telex, faxes, telegrams, and now through on-line computer, monitoring private and personal communications as they choose. In fact, the present day NSA is the current main expression of MJ-12 and PI-40 pertaining to the Flying Saucer Programs. Vast amounts of disinformation are spread throughout the UFO research field. Any witnesses to any aspect of the programs have their lives monitored in every detail, for each has signed a security catch. For people who have worked in the program, including military members, breaking that oath could have anyone of the following direct consequences:

- A verbal warning accompanied by a review of the security oath.
- A stronger warning, sometimes accompanied by a brow-beating and intimidation.
- Psychologically working on an individual to bring on depression that will lead to suicide.
- Murder of the person, which is made to appear as suicide or an accident.
- Strange and sudden accidents, always fatal.
- Confinement in a special Detention Center.
- Confinement in an Insane Asylum, where they are treated by mind-control and deprogramming techniques. The individuals are later released with changed personalities, identities, and altered memories.
- Bringing the individual into the inside, where he or she is employed and works for them and where he or she can be watched. This is usually in closed facilities with little contact with the outside world. Underground facilities are the usual place for this.

Any individual who they perceive to be too close to the truth, will be treated in the same manner. MJ-12/NSA will go to any length to preserve and protect the ultimate secret, as we will see later, the characteristics of what this ultimate secret would turn out to be would change drastically for it was something even MJ-12/NSA could not predict, actual contact with Alien groups. How the actual contact between the government and Aliens was initially made is not known at this time, but the Government was made aware that it could be done by an Alien using the right equipment.

# FACTS ABOUT ALIENS ON EARTH (I)
## CODE: OMNIDATA

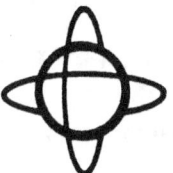

**IRS-28 - CODE: IRWING/7 - MAT/ORM, 26.01/AB/7**
After all these years of research we have compiled these major facts about Aliens on Earth.

- Alien crafts from other worlds have crashed on Earth.
- Alien crafts are from both Ultra-Dimensional sources and sources within this Dimension.
- Early U.S. Government efforts at acquiring technologies were successful.
- The U.S. Government has had live Alien hostages at some point in time.
- The U.S. Government has conducted autopsies on Alien cadavers.
- U.S. Intelligence Agencies and Security Agencies are involved in the cover-up of facts pertaining to the situation.
- People have been and are currently being abducted, mutilated, murdered and kidnapped as a result of the UFO situation.
- There is a current Alien presence on this planet among us that controls different elements of our society.
- Alien forces maintain bases on Earth and on the Moon.
- The U.S. Government has had a working relationship with Alien Forces for some time, with the express purpose of gaining technology in gravitational propulsion, beam weaponry and mind control.
- Millions of cattle have been killed in the process of acquiring biological materials. Both Aliens and the U.S. Government are responsible for mutilations, but for different reasons.
- We live in a multi-dimensional world that is overlapped and visited by Aliens/Entities from other dimensions. Many of those entities are hostile, and many are not hostile.
- The basis for our genetic development and religions lies in intervention by non-terrestrial and terrestrial forces.
- The truth about our actual technology far exceeds that perceived by the Public.
- The United States Space program of today is a cover operation that exists for public relations purposes.
- People are being actively killed in order to suppress the facts about the situation. The CIA and the NSA are involved so deeply that exposure would cause collapse of their overt structure.
- Facts indicate Alien presence with in 5 to 10 thousand years.
- Our civilization is one of many that have existed in the last billion years, "Milky Way time period".
- Alien Psychology (see information in the following pages).
- The Metagene factor, (see information in the following pages).

**MAIN IMPORTANT UFO CRASHES, UNDER ULTRA-SECRET COVER-UPS**

| NEW MEXICO, USA | 1947-1948 | CAUSE: Affected by a new experimental Radar Type |
| Roswell | | NOTE: This U.S. Radar is the basis to the |
| San Augustine Flats | | "JOSHUA and GABRIEL" defense projects. |
| Aztec | | |
| CALIFORNIA, Desert, USA | 1952 | CAUSE: Unknown |
| MEXICO, Sonora Desert | 1955 | CAUSE: Unknown |
| FLORIDA, Everglades, USA | 1959 | CAUSE: Affected by Radar. |
| TEXAS, Dayton, USA | 1961 | CAUSE: Unknown |
| BRAZIL, Mato Grosso | 1966 | CAUSE: Unknown |
| URUGUAY, Salto | 1970 | CAUSE: Unknown |
| BRAZIL, Amazon | 1976 | CAUSE: Unknown |
| MEXICO, Gulf of Mexico | 1979 | CAUSE: Unknown |
| ARGENTINA, Andes | 1979 | CAUSE: Unknown |
| ALASKA, Fairbanks, USA | 1981 | CAUSE: UFO Malfunction |

Of course there were other crashes, however, I was not sent on those investigations. Some of the other confirmed crashes occurred in Europe, Africa, China, Soviet Union, and Australia, just to name a few. With possible other un-confirmed crashes occurring in the Philippines and New Zealand.

**EXTRA-TERRESTRIAL CLASSIFICATIONS**

**GUESTS** — Invited Aliens, with authorization of the U.S. Government to stay on Earth.

**VISITORS** — Aliens who have come to earth for unknown missions.

**PRISONERS** — Captured Aliens by the U.S. Government and/or other Governments.

**RESEARCHERS** — Aliens who have come to Earth for Scientific Motives.

**INTRUDERS OR MALEVOLENTS** — Aliens who are dangerous for our civilization because they don't respect our society.

**NEUTRALS** — Aliens who just observe our civilization and never interfere with us.

**COLONIZERS** — Small groups of Aliens who have decided to live among us, (like us).

**INTERACTORS** — Small groups of Aliens who have decided to interfere with our History, making changes when its possible.

**ALIEN PRESENCE ON EARTH** - There are some one hundred sixty (160) or more known types of Aliens visiting our world (Earth) at the present time, these are the most commonly seen types:

**Greys, type one (1)** - The Rigelians from the Rigel Star system and are approximately four (4) feet tall, with a large head containing large slanted eyes, who worship technology and DON'T CARE ABOUT US. The type popularized in the "Communion" book by Strieber. They need vital secretions for their survival, which they are getting from us (earthlings).

**Greys, type two (2)** - Come from the Zeta Reticulae 1 & 2 solar systems. Same general appearance as a type one (1), although they have a different finger arrangement and a slightly different face. These Greys are more sophisticated then the type ones (1). They possess a degree of common sense and are somewhat passive. They don't require the secretions that the type ones (1) due.

**Greys, type three (3)** - Simple cloning form of types one and two above. Their lips are thinner (or no lips). They are subservient to the type one and two Greys above.

**Nordics, Blondes, Swedes** - Known by any of these names. They are similar to us. Blonde hair, blue eyes (some have dark hair and brown eyes and they're shorter in height). They will not break the law of non-interference to help us. They will only intervene if the Greys activity were to affect us directly.

**Nordic Clones** - They appear similar to us but with a grey tinge to their skin. These Nordics are controlled drones, created by the Greys, type ones (1).

**Intra-Dimensional (Not Para-Terrestrial)** - Entities that can assume a variety of shapes. Basically of a peaceful nature.

**Short Humanoids** - One and a half to two and a half feet tall, with skin bluish in color. They are seen quite frequently in Mexico near Chihuahua.

**Hairy Dwarfs (Orange)** - They are four (4) feet tall and weigh about thirty five (35) pounds. Their hair is the color of red. They seem to be neutral and respect intelligent life forms.

**Very Tall Race** - They look like us but are seven to eight feet tall. They are united with the Swedes.

**Men In Black (MIBs)** - They are not from the Delta or NRO division of the government. They are oriental or olive-skinned, there eyes are sensitive to light and have vertical pupils. They have very pale skin on some types. They do not conform easily to our social patterns. Usually they wear black clothes (sometimes all white or grey clothes), wear sunglasses and drive black cars. In groups they all dress alike. Sometimes time-disoriented. They cannot handle a psychological "curve ball" or interruption to their plans. They very often intimidate UFO witnesses and impersonate government officials. Equivalent of our CIA from another Galaxy.

# CATTLE MUTILATIONS

**General Chronology** - In the middle of 1963, a series of livestock attacks occurred in Haskell County, Texas. In a typical case, an Angus Bull was found with it's throat slashed and a saucer-sized wound in its stomach. The local population attributed the attacks to a wild beast of some sort, a so called vanishing varmint. As it continued throughout the Haskell County, the bloodluster assumed somewhat more mythic proportions and a new name destined to endure, "The Haskell Rascal".

Throughout the following decade, there would be sporadic reports of similar attacks on livestock. In 1967 the most prominent of these infrequent reports was the mutilation death of Snippy the horse in Southern Colorado, which at the same time were accompanied by accounts of UFOs and Unidentified Helicopters. In 1978, the attacks increased, and by 1979, numerous livestock mutilations were occurring in Canada, primarily in Alberta and Saskatchewan.

Attacks in the United States leveled off for a while, but in 1980, there was an increase in activity again. Since that year, mutilations have been reported less frequently, though this may be due in part to an increased reluctance to report mutilations on the part of ranchers and farmers. The mutilations still continue today, and over ten thousand animals have died in the United States alone, although the mutilations have been occurring worldwide, the circumstances surrounding the mutilations are always the same.

A drug connection to these mutilations may be found inside the "YELLOW BOOK". We found that, as the Terran Scientist learned about Longevity, (Secret of long life) the main basis for longevity was the capacity of human cells to recuperate. Anyone will grow old when their cells can't be restored and they start a process of deterioration and die. The secret of longevity is the restoration of the cells. This can be done using Altered Adrenalin, Alterated Coradrenalin, Cordrazyne or Cortropinex, (sometimes use only Formazinye and Hyronalinx, read "The Pulsar Project" for additional information on these and more durgs). All of these drugs have a base in Adrenalin, which is produced in the human brain. During the 60's the scientists discovered that they can be synthesized from the medullary portion of the adrenal gland of the cattle. They need big quantities in order to synthesize a small portion of the mentioned drugs above.

Of course, human scientist try to discover new applications and new synthesized drugs for recuperation of the cells, especially brain cells, and restoration to the human tissues and increasing the psychic and physical human skills.

**My General Personal Observations** - Any investigation which intends to probe the systematic occurrences of the mutilation attacks upon livestock and other animals must include within its preview certain factors which may or may not be directly related to the acts of mutilation themselves. These mutilations, the killing and furtive removal of external or internal parts, have been directed at literally thousands of animals, primarily livestock, since the 1960's. The surgery on these animals is primarily conducted with uncanny precision, suggesting the use of a highly sophisticated implements and techniques. The numbing and persistent regularity of the mutilations and the seemingly casual disposal of the useless carcasses all hints at extreme confidence, even arrogant's, of the mutilators. It is an arrogance which appears to be justified by the freedom and impunity with these acts have been carried out.

**My General Personal Observations - Human Reasons for the Mutilations** - These events, or the discussion of them, is just the precursor to the actual revelations of what is behind the mutilations: Alien acquisition of biological material for their own use. To discuss this in a logical and sequential manner, we must review what has been really happening right under our noses, "The direct interaction with Extra-Terrestrial Biological Entities (EBEs)". To discuss that, however, we must attempt to start at the beginning with what we know to be true.

**My General Personal Observations - Alien Reasons for the Mutilations** - The main interest to Aliens, (we talk specifically about Regelians) is breeding, with humans of course. They need the cattle's tissues because they have the same carron cells as the human being at Genetic levels. All incisions would be made by Replicas, Clones, or Androids. They can send teams and take the material they need, and it doesn't matter where or how much. If they need it, they just take it. This means that Rigelians are not Vampires, or that they need it for survival, but just immoral scientists that need material for experiments, (we really don't know what is worse).

**Black Helecopter Involvement In Mutilations** - The pertinence of a specific element of the problem is shortly revealed in the course of any thorough investigation into the mutilations. We refer to a appearance of unmarked and otherwise unidentified helicopters within a spatial and temporal proximity of animal mutilation sites. The occurrence of the two has been persistent enough to supersede coincidence. These mystery helicopters are almost always without identifying markings, or markings may appear to have been painted over or covered with something. The helicopters are frequently reported flying at abnormal, unsafe or illegal altitudes. They may shy away if witnesses or law officers try to approach.

There are several accounts of aggressive behavior on the part of the helicopter occupants, with witnesses chased, buzzed, hovered-over or even fired upon. At times these choppers appear very near mutilation sites, even hovering over a pasture where a mutilated carcass is later found. They may be observed shortly before or after mutilations occur, or within days of a mutilation. The intention here is merely to stress that the mystery helicopters did not develop concurrently with the animal mutilations themselves. Such helicopters, unmarked, flying at low levels, soundless, or sounding like helicopters, have been reported for years, and have been linked to an even more widespread phenomenon, the "Phantom" fixed wing aircraft.

As far as Tom Adams is concerned, the answer, could be a combination of the above explanations.

There also has been speculation that they are involved in biological experiments with chemical or biological warfare or the geobotanical pursuit of petroleum and mineral deposits. On one occasion, an Army Standard Type Scalpel was found at a mutilation site. Since the disks have been mostly involved with the mutilations, it is thought that this was a diversionary event.

In fact both, Alien and the U.S. Government are responsible for mutilations, but for different reasons.

# THE ANTAREAN CONNECTION
## OMNICORD

**Time and UFOs** - Time is one of the most important aspects of the UFO thing. It plays a strange, but significant role. Part of the answer may not lie in the Stars but in the clock ticking in front of you.

The planet Earth exists in three dimensions, we can move in many directions within these dimensions. Space does not exist except when we make it exist. To us, the distance between atoms in our matter is so minute that it can only be calculated with hypothetical measurements, yet, if we lived on a atom, and our size was relative to its size, the distance to the next atom would seem awesome.

There is another Earth's man-made measurement called time. Unlike the other three dimensions, time has us seemingly trapped. Time becomes very real to us, and it appears that we couldn't live without it. Yet time doesn't really exist at all. This moment exists to us, does this mean the same moment is being shared by other planets, in other galaxies ?, "not really".

The UFO phenomenon does seem to be controlled, this means it does follow intelligent patterns. If the objects themselves are manifestations of higher energies, then something has to manipulate those energies somehow and reduce them to the visible frequencies, for human patterns, but they take forms which seem physical and real to us, and they carry out actions which seem to be intelligent.

Thus, we arrive at the source, the source has to be a form of intelligent energy operating at the highest possible point of the Frequency Spectrum. If such an energy exists at all, it might permeate the Universe and maintain equal control of each component part. Because of its very high frequency, so high that the energy particles are virtually standing still, the source has no need to replenish itself in any way that would be acceptable to our environmental sciences. It could actually create and destroy matter by manipulating the lower energies. It would be timeless, because it exists beyond all time fields. It would be infinite because it is not confined by Three-Dimensional Space.

Perhaps if we were in a pure energy state, each particle of energy would itself serve as a synapse, and information could be stored by a slight alteration in frequency. All the memory fragments of a rose, for example, would be recorded at one frequency, and the whole energy form could tune into that memory by adjusting frequencies, as we might adjust a radio receiver, (well...well...well...where have we heard of this before!). In other words, no complex circuitry would be required, and no physical body would be necessary either. The energy patterns would not need material form, it instead would permeate the entire Universe, (Universe,.... interesting small point of view of a gigantic concept of the reality of the MULTIVERSE). It could surround you completely at this very moment and be aware of all the feeble impulses of low energy passing through your brain. If it were so desired, it would control those pulses and thus control your thoughts. Man has always been aware of this intelligent energy or force, he has always worshipped it, i.e. God, Jesus, or the Universal Being, etc.

1. Our first conclusion is that some UFOs originate from beyond our own time frame or time cycle.

2. Our second conclusion is that the source has total foreknowledge of human events and even individual lives, since time and space are not absolutes.

These two conclusions are compatible.

It is that all human events occur simultaneously when viewed by a greater intelligence. If a greater intelligence wants to communicate with a lower form, all kinds of problems are presented. The communication must be conducted in a manner which will be meaningful and understandable to the lower life form, an acceptable frame of reference must be found and utilized.

The UFO phenomenon is frequently reflective; that is, the observed manifestations seem to be deliberately tailored and adjusted to the individual beliefs and attitudes of the witnesses. Contactees are given information which, in most cases, conforms to their beliefs. UFO researchers who concentrate on one particular aspect or theory find themselves inundated with seemingly reliable reports which tend to substantiate that theory, (?!).

Soviet researchers extensive experiences with this reflective factor led them to carry out weird experiments which confirmed that a large part of the reported data is engineered and deliberately false. The witnesses are not the perpetrators, but merely the victims.

The apparent purpose of all this false data is multifold, much of this is meant to create confusion and diversion. Some of it has served to support certain beliefs which were erroneous but which would serve as a stepping stone to the higher, more complex truth. Whole generations have come and gone, happily believing in the false data, unaware that they were mere links in the chain.

If it were all understood too soon, we might crumble under the weight of the truth. The planet Earth is covered with windows, (corridors, tunnels, and/or gates) into those other unseen worlds. We have the instruments to detect them, and if we decide to use them, we would find that these windows are the focal point for Super High-Frequency Waves, the rays of ancient lore. These rays might come from Orion or the Pleiades Star System as the Terran's ancients claimed, or they might be part of the great force that emanates throughout this Universe, perhaps from the Antares Star System. The UFOs have given to the Terrans the evidence that such rays exist. Now, slowly, we are being told why.

Observers from the Antares Star System are watching the Terrans and they know the guidelines for "Warp Zones" come from the center of the Antares Star System and draw tunnels for several main points of this and other Universes. They are "Gates" for several travelers from all Universes. Of course there are many more focus points, when other rays come from around the Universe.

# CROP CIRCLES PHENOMENA
## CODE: SEC

**Crop Phenomena: Circles Affair**

The better way to understand the so called crop phenomena, (circles that appear in fields) is by the following example:

We, the people of Earth, exist at the 3rd dimension because our atoms have a specific frequency which makes us able to exist at the 3rd dimension. This specific frequency is stable enough for all our lifetime. One quick idea of dimensions tells us:

1st Dimension is just a plain point.

2nd Dimension is two points together connected by a line.

3rd Dimension is Density, Volume (us and our universe)

4th Dimension ———————————————— TIME

Our main interest in this report is the:

5th Dimension ———————————————— Parallel Universes or Multiverse

That's the whole idea behind the multiverse concept, not just one universe but an infinite number of parallel universes. Using math to make a simplification of this concept tells us that there are no limits of the higher or lower number that exist. It's impossible to figure these numbers like it's impossible to figure the extension of possibilities of the Multiverse.

If we are capable of accelerating or decelerating the frequencies to make us able to exist in the 3rd Dimension, we can jump to the 5th Dimension, or the Multiverse.

To understand this idea better, try to think that all this is just like listening to a radio set. When you listen to a radio set, you are listening to a single radio station, on a single frequency, by changing to a different radio station, you have to change to a different frequency. This is accomplished by the frequency tuning dial on the front of the radio set, but because you are listening to a single radio station, on a single frequency, does not mean that other stations on other frequencies do not exist, right.

Same thing with the other universes from the Multiverse, they're at other different frequencies than we are and we just don't know it yet. Why are we in the 3rd Dimension and they are in the 5th Dimension?, why not the reverse?, simple, "its based on perspective, they see us as their 5th Dimension and we see them as our 5th Dimension". Again, it is just a question of perspective.

The other dimensional travellers come to our dimension by using some process of acceleration of particles and give them the capability of jumping between dimensions, but why did they select England first, and then why the fields ?

The answer is England, had all of the right conditions for one of several civilizations that exist in the Multiverse to be able to contact us through crop circles. All answers of how they were able to accomplish this task lie in a vast prehistoric energy circuit.

England has a complex web of energy running in inter-links with tumull, ancient graves and stone circles or henges, and all of the sites, henges, mounds and stone circles were constructed in relation to underground water linked with energy lines. This creates some sort of natural electro-magnetic force field created by the water, combined with the presence of underground quartz deposits. Also, the lunar gravitational effects on the water, land mass and the build up of a electrical magnetic charge from the quartz produces a periodic energy discharge.

The energy bolts give the necessary conditions for these other dimensional civilizations to be able to start experimenting with our universe and/or planet.

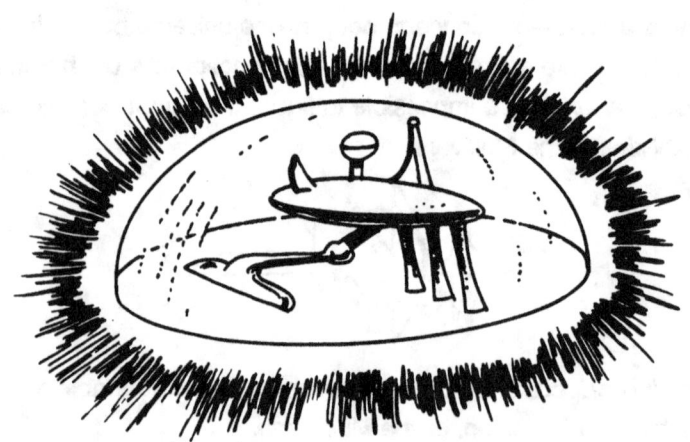

They have started to send probes here, in order to know more about the natural conditions or our universe and planet. The probes are limited to one specific semidome of energy, a circle, in a field where they have scanned the energy bolt.

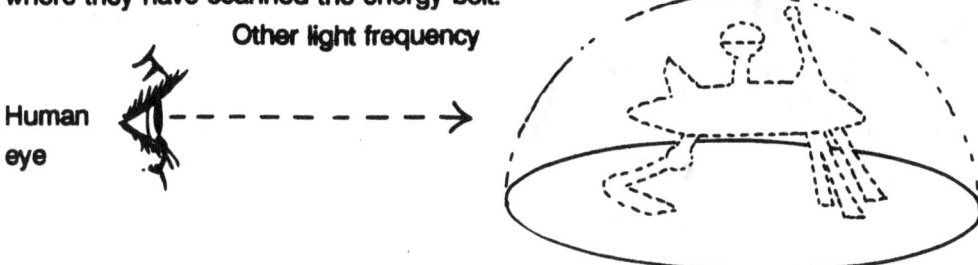

In the above sketch, our eyes wouldn't be able to see the probe at that moment, because of our visual limitation, of not being able to see the frequencies between the frequencies of infrared and ultraviolet. Of course if you were able to detect the device using infrared or ultraviolet equipment, then you would be able to see it.

The crop circle phenomena didn't start now, or even a few decades ago, it started centuries ago and just got more and more complex. Normally you detect the proximity of a circle by ionization of the air near you, i.e. changes of the temperature and/or a characteristic sound produced by the acceleration of particles, when they are semi-stable at our reality.

They are now experimenting all over the planet. Always looking for similar conditions to England to send messages, "very particular messages", using domes, which are fluxes of particles from their particle accelerator device.

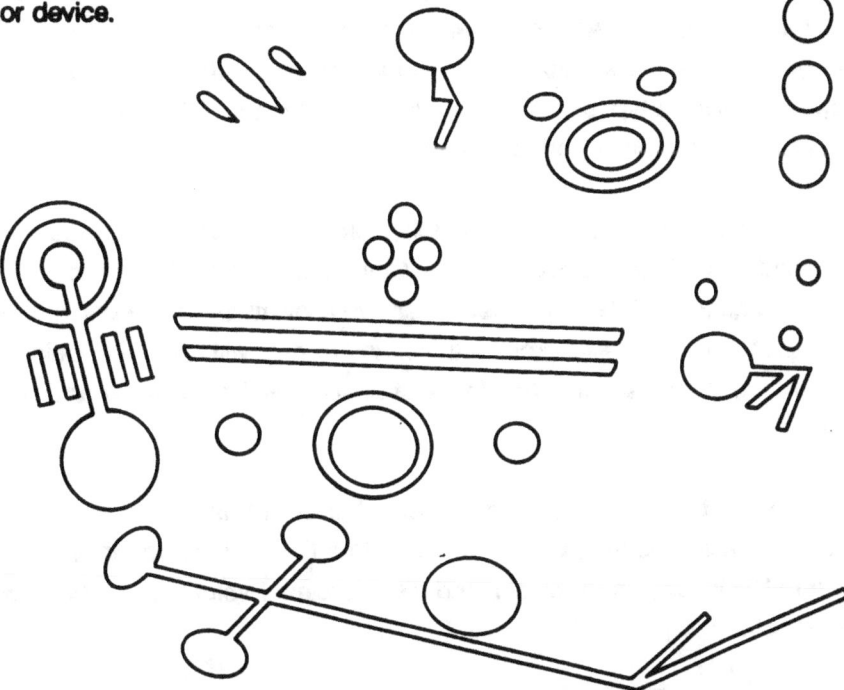

Very soon they will be able to make a total materialization and real existence at our reality.

# THE DULCE BASE
## CODE: J.B. III

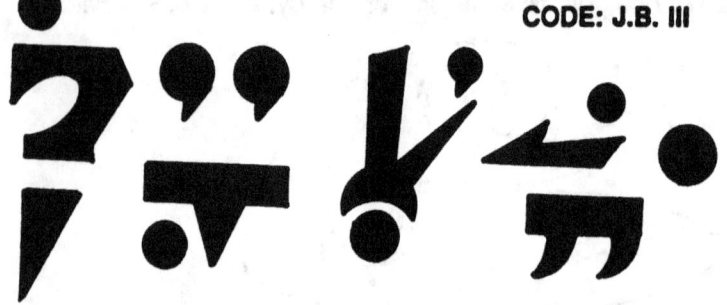

The following material comes from people who know the Dulce underground base exists. They are people who worked in the labs; abductees taken to the base; people who assisted in the construction; intelligence personnel, (NSA, CIA; etc.) and some specific UFO inner Earth researchers. This information is meant for those who are seriously interested in the Dulce base. For your own protection, be advised to "Use Caution" while investigating this complex. An ongoing investigation made by J.B. III, (Code: SR24.3B7).

**The Dulce Base**
This facility is a genetics lab and is connected to Los Alamos, via Tube-Shuttle. Part of their research is related to the genetic effects of radiation, (Mutation and Human Genetics). Its research also includes other Intelligent Species as well, (Alien Biological Life Form "Entities"). In the revised September 1950 edition of The Effects of Atomic Weapons, prepared for and in cooperation with the U.S. Department of Defence and The U.S. Atomic Energy Commission, under the direction of the Los Alamos Scientific Laboratory, we read about how complete underground placement of bases is desirable. On page #381: "There are apparently no fundamental difficulties in construction and operating underground various types of important facilities. Such facilities may be placed in a suitable existing mine or a site may be excavated for the purpose".

**Caught in the Game** - Centuries ago, surface people, some say the Illuminati, entered into a pact with an "Alien Nation", hidden within the Earth. The U.S. Government, in 1933, agreed to trade animals and humans in exchange for high-tech knowledge, and to allow them to use undisturbed underground bases, in the western U.S.A. A special group was formed to deal with the Alien beings. In the 1940's, "Alien Life Forms, (ALFs)" began shifting their focus of operations, from Central and South America to the United States.

The continental divide is vital to these specific entities. Part of this has to do with magnetic substrata rock and high energy plasma states, (see: Beyond the Four Dimensions, reconciling physics, parapsychology and UFOs, by Karl Brunstein, and also: Nuclear Evolution, discovery of the rainbow bodi/aura by Christopher Hills).

- This area has a very high concentration of lightning activity; underground waterways and cavern systems, fields of atmospheric IONS, etc.

**Who's Planet is this ?** - These specific Aliens consider themselves Native Terrans. They are an Ancient race, decedents of a reptilian Humanoid species which cross-bred with some of the more primitive Uni-Terrestrial survivors. They are untrustworthy manipulator mercenary agents from another Extra-Terrestrial culture, "The Draco's" who are returning to Earth, which was their ancient outpost before the coming of the original Uni-Terrestrials, to try to use it as a staging area, which is not easy at all, because it causes all the other one hundred seventy (170) different Alien species to want their share of the Metagene secrets. But, these Alien cultures are in conflict over who's agenda will be followed for this planet. All the while mind control is being used to keep humans in place, artificially of course, especially since the forties. The Dulce complex is a joint U.S. Government and Alien base. It was not the first one built with the Aliens and others are located in Colorado, Nevada, Arizona, Alaska, etc.

**The Secret Activity** - About Dulce area: Troops went in and out of there every summer, starting in 1947. The natives do recall that they also built a road, right in front of the people of Dulce and trucks went in and out for a long period. That road was later blocked and destroyed. The signs on the trucks were "Smith Corp." out of Paragosa Springs, Colorado. No such corporation exists now, no records exist, and I believe the base, at least the first one was being built then undercover of a lumbering project, who never hauled any logs at all, only big equipment.

**R & D and the Military Industrial Complex** - The Rand Corp. became involved by doing a study, for the base on geology around the near by lakes. Most of the lakes near Dulce were made, VIA Government Grants, "for" the Indians, as such a completed Grant, the NAVAJO Dam is a main source for conventional electrical power, with the second source in ELVADO (which is also, an underground entrance to the Dulce base).

NOTE: If Rand is the mother of "Think Tanks", then the "Ford Foundation" must be considered the Father.

Rands secrecy is not confined to reports, but on occasion extends to conferences and meetings. On page #645 of the Project Rand, proceedings of the Deep Underground Construction Symposium of March 1959, we read:

"Just as airplanes, ships and automobiles have given man mastery of the surface of the Earth, tunnel, boring machines, will give him access to the subterranean world."

NOTE: **The September 1983 issue of OMNI, page 80,** has a color drawing of "The Subterrene", the Los Alamos Nuclear powered tunnel machine, that burrow through the rock, deep underground, by heating whatever stone it encounters into molten rock, (Magma), which cools after the subterrene machine has moved on. The result is a tunnel with a smooth, glazed lining. These underground tubes are used by electro-magnetically powered sub-shuttle vehicles, which can travel at great speeds. They connect the so-called "Hidden

Empire" sub-city complexes. Also, the top secret project code-named "NOAH'S ARK", uses the tube shuttles in connection with a system of over one hundred (100) bunkers and bolt holes, which have been established at various places on Earth. They built the same type of subterranean tunnels at the ultra top secret Moon and Mars bases as well. Many of these underground cities are complete with streets, sidewalks, lakes, small electrical cars, buildings, offices and shopping malls.

There were over six hundred fifty (650) attendees to the 1959 Rand Symposium, most were representatives of the Corporate Industrial State, like:

- The General Electric Company
- Hughes Aircraft Company
- Sandia Corporation
- Walsh Construction Company
- Colorado School of Mines

- AT&T
- Northrop Corporation
- Stanford Research Institute
- The Bechtel Corporation
- Etc.

Bechtel (pronounced Beck-tul) is a super secret international corporation octopus, founded in 1898. Some say the firm now is really a "Shadow Government", a working arm of the CIA. It is the largest construction and engineering outfit, in the U.S.A., the world, and some say, beyond. The most important posts in U.S.A. Government are held by former Bechtel officers. They are part of "The Web", an inter-connected control system, which links the Tri-Lateralist Plans, the C.F.R., the Orders of Illuminism, (cult of the All-Seeing Eye) and other interlocking groups.

**Surviving The Future** - The Dulce facility consists of a central hub, the security section, and also some photo labs. The deeper you go, the stronger the security. This is a Multi-Leveled complex, with over 3,000 cameras at various high security locations, i.e. exits and labs.

There are over one hundred (100) secret exits near and around Dulce. Many around Archuleta Mesa, others to the south around Dulce lake and even as far East as Lindrith.

NOTE: Deep sections of the complex connect into natural cavern system. A person who worked at the Base (CR-24/ZM 35-File IV), who had an "Ultra 7-B" clearance reports the following:

"There may be more than seven levels, but I only know of seven. Most of the Aliens are on levels 5, 6, and 7. Alien housing is on level five (5)."

**Transamerican Underground Subshuttle System, (T.A.U.S.S.)**

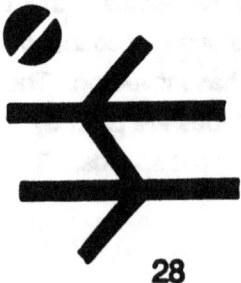

**Overt and Covert Research within Dulce** - As U.S. Energy Secretary, John Herrington named the Lawrence Berkely Laboratory and New Mexico's Los Alamos National Laboratory to house new advanced genetic research centers as part of a project to decipher the Human Genome. The Genome holds the genetically coded instructions that guide the transformation of a single cell, a fertilized egg, into a biological being. "The Human Genome Project may well have the greatest direct impact on humanity of any scientific initiative before us today", said David Shirley, Director of the Berkeley Laboratory.

Covertly, this research has been going on for years, at Dulce labs. Level #6, at Dulce, is privately called "Night Mare Hall", it holds the Genetic Labs. Reports from workers, (CR-24/ZM 52-Files VII), who have seen bizarre experimentation, are as follows:

> "I have seen multi-legged humans that look like half-human, half-octopus. Also reptilian-humans, and furry creatures that have hands like humans and cries like a baby, it mimics human words, and also huge mixture of lizard humans in cages."

There are fish, seals, birds and mice that can barely be considered those species. There are several cages, and vats of winged humanoids, grotesque Bat like creatures about three and a half to seven feet tall. Gargoyle-like beings and Draco-Reptoids.

Level #7 is worse, row after row of several humans and humanoids, (hundreds, perhaps thousands) in cold storage. Here too are embryo storage vats of humanoids, in various stages of development.

> "I frequently encountered humans in cages, usually dazed or drugged, but sometimes they cried and begged for help. We were told they were hopelessly insane and involved in high risk drug tests to cure insanity. At the beginning, we were told never try to speak to them at all at the and we believed that story."

Finally in 1978, a small group of workers discovered the truth. It began the Dulce Wars, and a secret resistance Unit was formed.

    **NOTE:** There are over 18,000 Aliens at the Dulce complex at this time.

In late 1979, there was a confrontation, over weapons, a lot of scientists and military personnel were killed. The base was closed for a while, but now, it is currently active again.

**NOTES:** 1. Human and animal abductions, for their blood and other parts, slowed in the mid-1980's, when the Livermore Berkeley Labs began production of artificial blood for Dulce and its sister complexes.

    2. About the confrontation, Human and Alien at Dulce, eighty (82) people were killed between

scientists and the National Recon Group, the DELTA GROUP, which is responsible for security of all Alien connected projects. Also, there were hundreds of other wounded people and one hundred thirty two (132) dead Aliens.

This type of entity is one that abductees and contactees have seen in underground breeding facilities since 1963. They are bred by the thousands in underground facilities, the Dulce facility is the most well known place that this activity occurs, although there are at least another twenty six (26) bases within the United States that have similar facilities.

The Delta Group, within the Intelligence Support Activity, have been seen with badges which have a black triangle on a red background. A Delta is the fourth letter of the Greek alphabet, it has the form of a triangle, and figures prominently in certain Masonic Signs.

**Each Base Has Its Own Symbol** - The Dulce Base symbol is a triangle with the Greek letter "TAU", (T) within it and then the symbol is inverted, so the triangle points down.

The insignia of "a triangle and three (3) lateral lines" has been seen on Alien Transport Craft, The Tri-Lateral Symbol.

Other symbols mark landing sites and Alien Craft.

**Inside the Dulce Base** - Security officers wear jumpsuites, with the Dulce Symbol on the front upper left side. The standard hand weapon at Dulce, is a "Flash Gun", which is good against human and Aliens. They use a Retina-Reader for Identification in substitution of the old ID Cards, but the ID Card readers are still used in card slots, for the doors and elevators, which had the Dulce Symbol above the ID Photo. After the second level, everyone is weighed in the nude, then given a uniform, visitors are given one off-white uniform. In front of all sensitive areas are the Retina-Readers and scales built under the doorway, by the door control. The clearance is checked, identity is checked and also the weight. Any possible change over two pounds and a security yellow is alerted.

No one, without a specific clearance or authorization, is allowed to carry anything into or out of sensitive areas. All supplies are put through a security conveyor system. The Alien symbol language, Grey and Nordic appear a lot at the facility.

During the construction of the facility, which was done in stages, over many years, the Aliens assisted in the design and construction materials. Many of the things assembled by the workers were of a technology they could not understand, yet, now a days it is a different story, it would function when fully put together. Example:

> The elevators have no cable, they are controlled magnetically. The magnet system is inside the walls. There are no conventional electrical controls. All is controlled by advanced magnetics. That includes a magnetically induced, (phosphorescent), illumination system. There are no regular light bulbs, and all exits are magnetically controlled.
>
> Note: If you were to place a large electromagnet on an entrance, it will affect an immediate interruption. They will have to come and reset the system, and "blow away your head, if they catch you !"

**The Town of Dulce** - The area around Dulce has had a high number of reported animal mutilations. The Government and the Aliens used the animals for environmental tests, psychological warfare on people, etc. The Aliens also wanted large amounts of organic material for Genetic Research, Nutritional and other reasons.

In the book, "ETs and UFOs, They need us, we don't need them", was an original idea for one title, but not my personal choice, by Virgil "Posty" Armstrong, he reports how his friends Bob and Sharon stopped for the night in Dulce and went out to dinner, they overheard some local residents openly and voicefully discussing Extra-Terrestrial abductions of the townspeople for purposes of experimentation. The ETs were taking unwilling human guinea pigs from the general populace of Dulce and implanting devices in their heads and bodies. The townspeople were frightened and angry but didn't feel they had any recourse since the ETs had our Government's knowledge and approval, (CODE: SR-24/AK.5).

Recently, participants in a field investigation, (CODE: SR-24/R25/AK.2), of the area near Archuletta Mesa, were confronted by two small hovering spheres. They all became suddenly ill and had to leave the area. Several Dulce residents are not naturally from this area, single people and even couples with children, have just come to the town of Dulce since 1948.

Generations of frozen agents occupied in suspect positions like at work in a gas station, drugstore, bar, restaurant, etc. They're there to listen and report anything which violates their limit of security. Always in the town of Dulce, you never know who is who !

**Mind Manipulation Experiments** - The Dulce Base has studied mind control implants; Bio-PSI units; ELF Devices capable of mood, Sleep and Heartbeat control; etc.

Defense Advanced Research Projects Agency, (DARPA) is using these technologies to manipulate people. They establish "The Projects", set priorities, coordinate efforts and guide the many participants in these undertakings. Related Projects are studied at Sandia Base by the "Jason Group" of fifty five scientists from thirty eight specific scientific areas. They have secretly harnessed the Dark Side of technology and hidden the beneficial technology from the public. Other projects take place at "Area #51" at Groom Lake in Nevada, codename "Dreamland", which is a data and ongoing projects repository and establishment for just some of the following:

- Electro-Magnetic Intelligence, (ELMINT)
- Code: Empire
- Code: EVA
- Prometheus Project
- Hybrid Intelligence System Program, (HIS)
- BW/CW
- Infrared Intruder Systems Project, (IRIS)
- BI-PASS
- REP-TILES

The studies on Level #4, at Dulce, include human/Alien, Aura Research, as well as, all aspects of Dream, Hypnosis, Telepathy, etc. They know how to manipulate the Bioplasmic body of humans. They can lower your heartbeat with deep sleep "Delta Waves", induce a static shock, then re-program via a brain-computer link. They can introduce data and programmed reactions into your mind, (information impregnation, "The Dream Library"). We are entering an ERA of the technologicalization of psychic powers, this means, the development of techniques to enhance man and machine communications;

- Nano-Technology
- Bio-Technology
- Micro-Machines
- PSI-War
- Electronic Dissolution of Memory, (EDOM)
- Radio-Hypnotic Intra-Cerebral Control, (RHIC)

as well as, various forms of behavior control via;

- Chemical Agents
- Ultra-Sonics
- Optical Frequencies
- Electro-Magnetic, (EM) Frequencies

which are directed to the physics of "Consciousness".

**Better Living Through Bio-Tech ?** - The development of "Bio-Technologies" will mean a revolutionary change in life of every human being now on Earth ! **WARNING ! "FASCISM IS CORPORATISM"**. We have passed the point of no return, in our interaction with the Alien beings. We are guaranteed "A CRISIS" which will persist until the final revelation or conflict.

The crisis is Here, Global and Real. We must mitigate or transform the nature of the disasters to come, and "come they will". Knowing is half the battle.

<u>ADVICE</u>: Read the Book, "The Cosmic Conspiracy" by Stan Deyo.

**The Phantom Empire: "Above the Law"** - The Dulce Base is run by a Board:
- The Chairman of the Board is John Herrignton.
- Jim Baker of Tennessee is the NSA/CIA link to Dulce.
- House speaker Jim Wright of Dallas Texas, the Nation's third highest office, is the Treasurer at Dulce.

There is currently a power struggle going on, as Rep. William Thomas, R-California, put it, "Part of Jim Wright's problem is he fails to understand what's equitable and fair. It's the arrogance of power". Even among his fellow Democrats, many find wright to be uncomfortable, Wright's operating style leaves him vulnerable.

Most meetings of the Dulce Board are held in Denver, Colorado and Taos, New Mexico. Former New Mexico Senator Harrison, publicly known only as "Last Man on the Moon" Schmitt has full knowledge of Dulce. He was one of seven astronauts to tour the base. In 1979, he held a "Animal Mutilation" conference in Albuquerque, New Mexico. This was used to locate researchers and determine what they had learned about the links between the "Mute" operations and the Alien/Government deals.

Senator Brian of Nevada, knows about the "Ultra Secrets at Dreamland" and Dulce. So do many others in the Government, "this is what the UFO researchers are up against", SO BE CAREFUL ! *They have killed to keep this information secret, and by reading this document, you now know more than they want you to know !*

They also have underwater bases off the cost of Florida and Peru. More detailed information will be released by the Government following the plan of preparation for a Global Alien revelation, in the very near future; Photo's, Video Tapes, Documents, etc. will be released. Watch for Agents among you now.

After all, a fascist group, within this country, had John Kennedy assassinated and got away with it. Look to the links, within the larger umbrella, the "Web" of a fascist totalitarian Secret Police State, within the Pentagon; JCS; DIA; Division Five of the FBI; DISC/DIS and CIA.

NOTE: The Defensive Investigative Services, (DIS) - Their insignia is a composite of the Sun's Rays, a Rose, and a Dagger, Symbolizing "The search for information, trustworthiness and danger".

**The following are basic diagrams of process of creation of artificial hybrids at level #7, at the Dulce Complex:**

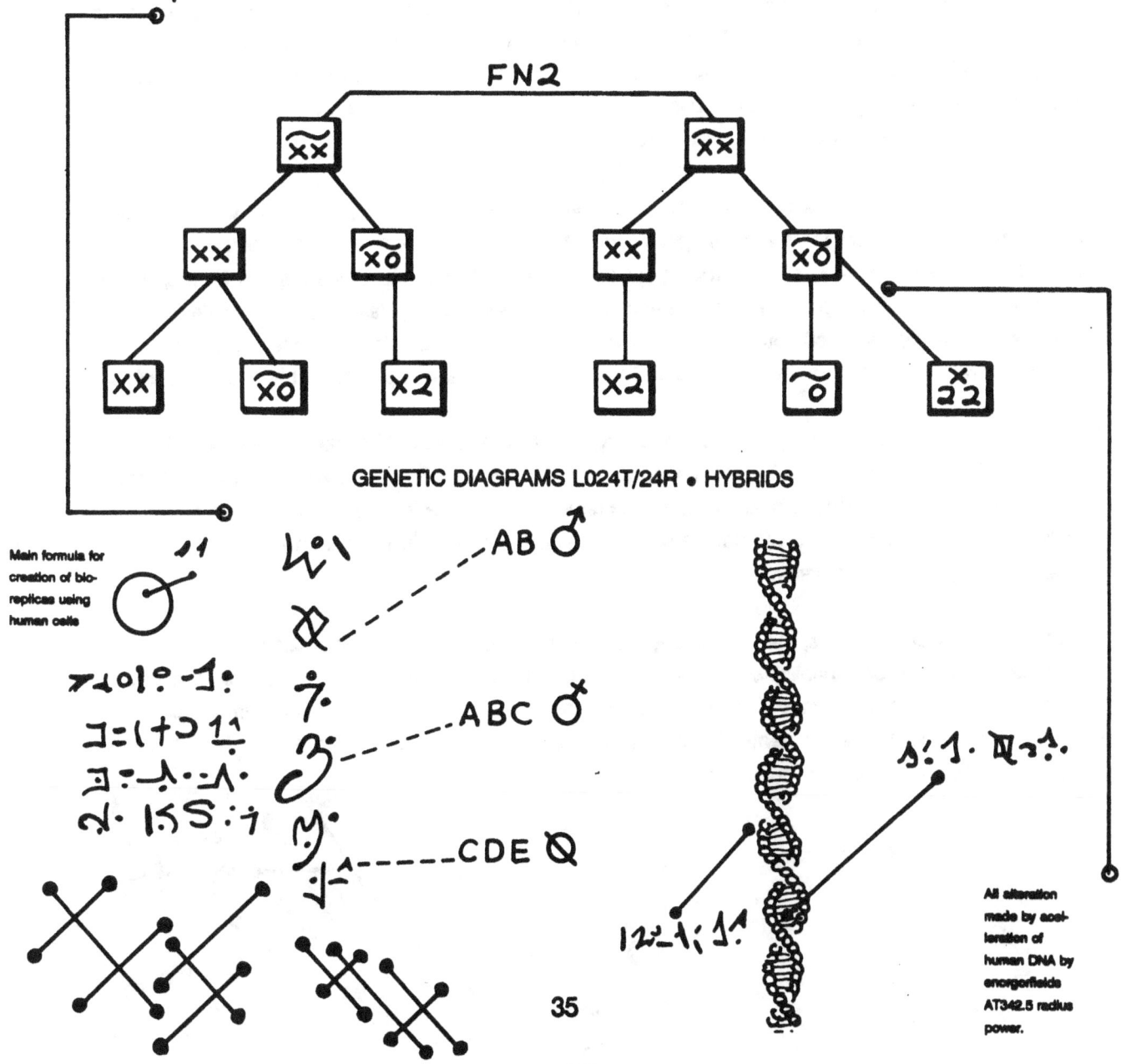

# SOME KINDS OF ALIEN LIFEFORMS WE KNOW ABOUT

**THE U.S. GOVERNMENT HAS CONDUCTED AUTOPSIES ON ALIEN CADAVERS (Grey Physiology and Anatomy)** - The most significant Alien specimens, which have been available to autopsy have been the "Grey Types".

The approximate height of most specimens are between 3.5 and 4.5 feet. The head, by human standards, is larger in comparison with the body. Facial features show a pair of eyes described as large, sunken or deeply set, far apart or distended more than a humans, and slightly slanted as oriental or mongoloid. No ear lobes or apertures on the side of the head were visible. The nose is vague, one or two small holes have been identified as nostrils. The mouth area is described as a small slit or fissure. In some cases, there is no mouth at all. It appears not to function as a means for communication or for alimentation. The neck area is described as being thin, and in some cases, not being visible at all because of the tightly knit garment. Most observers describe these humanoids as being hairless. Some of the bodies recovered have a slight hair patch on top of the head. Others appear with what appears to be like a silver skullcap.

There were no breathing attachments or communication devices. This suggests telepathy with higher intelligence. In one case, *there was an opening in the right frontal lobe area, revealing a crystalline network.* This network implies the development of some sort of third brain. *We also found a spherical object which was more or less 1 cm in size and was linked with the crystalline network.* We theorize that this is a device for amplification of their brain waves, which means they are not capable of psychic activity without the spherical device. This information is also included in the "YELLOW BOOK".

The arms are described as long and thin, reaching down to the knees. The palms each contained four fingers, with no thumbs. Three fingers are longer than the others. Some are very long, and others are very short. No description is available for the legs and feet. Some pathologists indicate that the section of the body was not developed as we would anticipate, showing that some of these beings were adapted to life in the water.

According to most observers, the skin is grey. Some claim it is beige, tan, or pinkish grey. No reproductive organs or capabilities were discovered, no phallus or womb. This suggests cloning is the current reproductive method, which has been mentioned in other sources. The humanoids appear to be a mold, sharing identical racial and biological characteristics. There is no blood as we know it, but there is a fluid which is grayish in color.

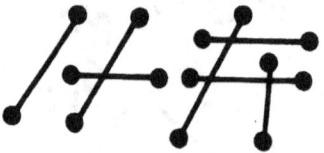

**UNI-TERRESTRIAL - CODE: UNA** (Seven ancestral Alien races that started our civilization here). Some of the unanswered questions about the human evolution, started centuries ago, when disaster struck a mission involving seven different Alien races to Earth. This information was recovered from **PROJECT GENESIS III (G-III) - ADN6.2 - CR-7/26TSW-3** and **CLR-25/M6-722 - CLAS.ATR26/AC #672/B25**, as well as other crash sites.

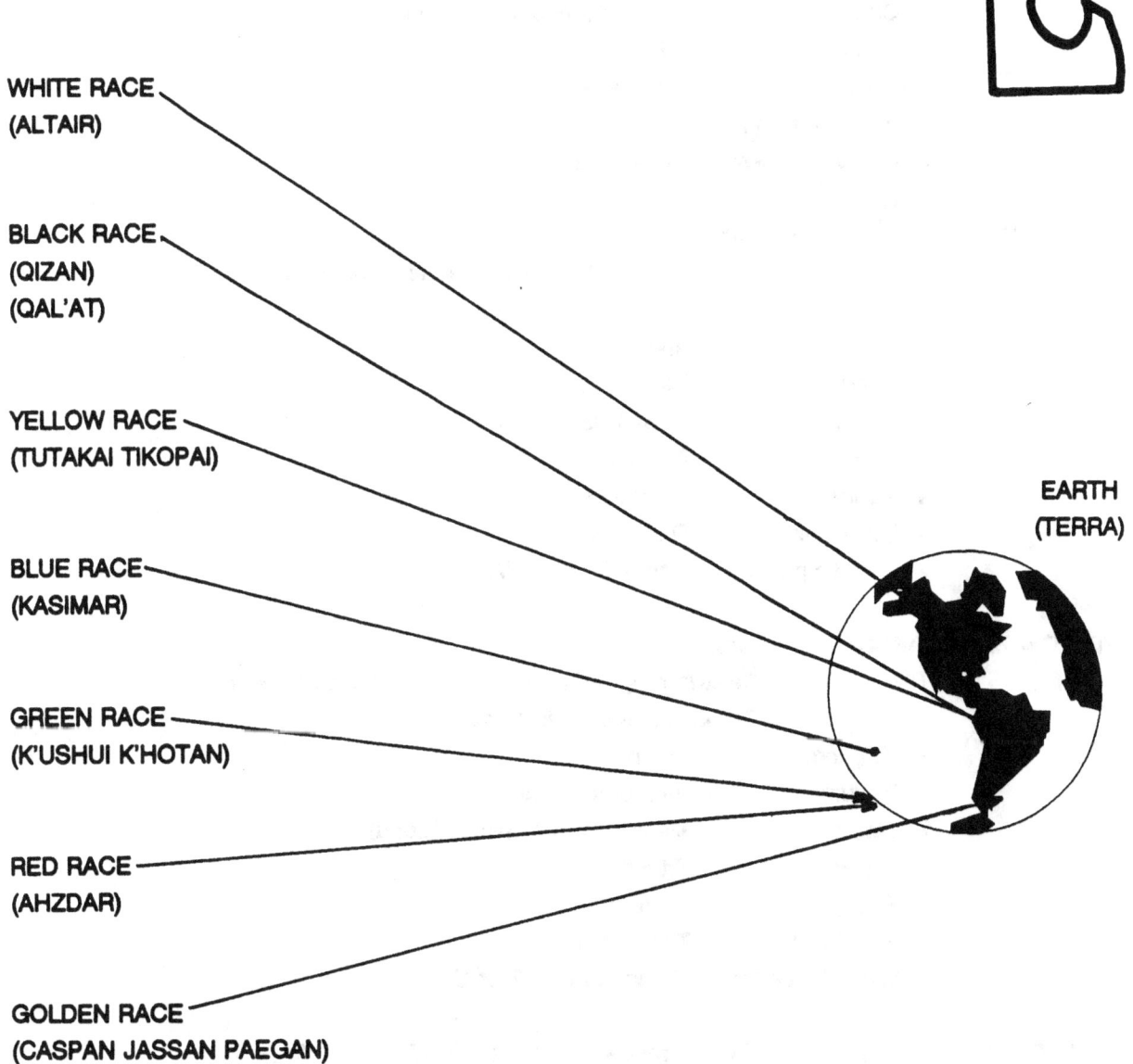

WHITE RACE (ALTAIR)

BLACK RACE (QIZAN) (QAL'AT)

YELLOW RACE (TUTAKAI TIKOPAI)

BLUE RACE (KASIMAR)

GREEN RACE (K'USHUI K'HOTAN)

RED RACE (AHZDAR)

GOLDEN RACE (CASPAN JASSAN PAEGAN)

EARTH (TERRA)

## ANTHROPOLOGICAL DESCRIPTION OF EACH RACE OF ALIENS

**WHITE RACE**    **Altair**    - Relative range of absolute visual magnitude (+2).
- Carbon Hominoid lifeform.
- Kingdom: Animal
- Phylum: Vertebrata (Obs)
- Class: Mammalia (Hominid Biped)
- Order: Primata
- Family: Hominidae
- (no sub-family)
- Genus/Species: Homo Altair

**BLACK RACE**    **Qizan Qal'At** - Binarie
- Relative range of absolute visual magnitude (+7E)
- Carbon Humanoid lifeform.
- Kingdom: Animal
- Phylum: Vertebrata
- Class: Mammalia
- Order: Primata
- Family: Hominidae
- Sub-Family: Qal'Atzoid
- Genus/Species: Homo Qizan Qal'At

**YELLOW RACE**    **Tutakai**    - Binarie
**Tikopai**    - Relative range of absolute visual magnitude (+7E)
- Carbon Humanoid lifeform.
- Kingdom: Animal
- Phylum: Vertebrata (Obs)
- Class: Mammalia (Humanoid Biped)
- Order: Primata
- Family: Hominidae
- Sub-Family: Tikopaizoid
- Genus/Species: Homo Tutakai Tikopai

**BLUE RACE**    **Kasimar**    - Relative range of absolute visual magnitude (+5)
- Carbon Humanlife lifeform.
- Kingdom: Animal
- Phylum: Vertebrata (Obs)
- Class: Mammalia (Humanoid Biped)
- Order: Primata (Anphibio)
- Family: Hominidae

- Sub-Family: Mermannus Kasimarzoid
- Genus/Species: Homo Kasimar

**GREEN RACE** K'ushui   - Binarie
K'hotan   - Relative range of absolute visual magnitude (+7E)
  - Carbon Humanoid lifeform.
- Kingdom: Animal
- Phylum: Vertebrata (Obs)
- Class: Mammalia (Humanoid Biped)
- Order: Primata
- Family: Hominidae
- Sub-Family: K'hotanzoid
- Genus/Species: Homo K'ushui K'hotan

**RED RACE**   Ahzdar   - Relative range of absolute visual magnitude (+5)
  - Carbon Humanoid lifeform.
- Kingdom: Animal
- Phylum: Vertebrata (Obs)
- Class: Mammalia (Humanoid Biped)
- Order: Primata
- Family: Hominidae
- Sub-Family: Hominahzdarzoid
- Genus/Species: Homo Ahzdar

**GOLDEN RACE** Capsan   - Trinarie
Jassan   - Relative range of absolute visual magnitude (+7E)
Paegan   - Carbon Humanlife lifeform.
- Kingdom: Animal
- Phylum: Vertebrata (Obs)
- Class: Mammalia (Humanoid Biped)
- Order: Primata
- Family: Hominidae
- (no sub-family)
- Genus/Species: Homo Caspan Jassan Paegan

## MISSION SKILLS & ASSIGNMENTS

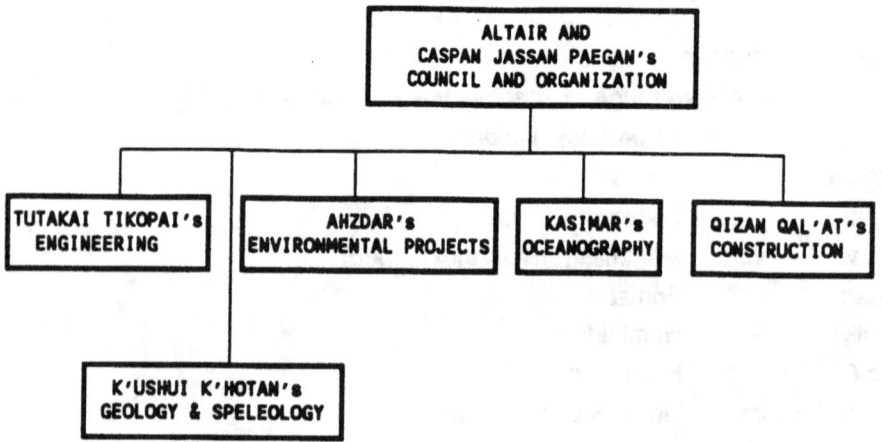

**MISSION:** Genetic Research and Colonization of the 3rd Planet (Terra) of the Star (Sol) "our sun" of the Milkyway Galaxy (N'erandha B'hai).

**STATUS:** OUT TO CONTROL.

**GROUP:** Altaires, Qizan Qal'Ats, Tutakai Tikopais, Kasimars, K'ushui K'hotans, Ahzdars, Caspan Jassan Paegans.

**SEQUENCES OF MISSION:**

a. Division of the area looks more like their original planets. Division of Terra, Mars, and Phaeton.
b. Research with Earth Reptilians - caused a mutation now called the dinosaurs.
c. Research with Mammalia mutations.
d. Erased (Killed Off) Dinosaurs.
e. Crisis at Star Sol, (our Sun).
f. Destruction of the original mission.
g. Survivors and their mental problems, left from destruction of mission.
h. War between survivors.
i. Destruction of Phaeton, creation of asteroid belt, (Van Allen asteroid belt).
j. Survivors.
k. Survivors separated and created the Ancient civilizations known as Lemuria, Mu, and Atlantis.
l. Homo Sapiens, Homo Mermanus, and Homo Interior co-exist on Earth with survivors.
m. New Aliens Landings by the Dominium Grey and Nordic races.

**MISSION
STAR MAP
Code: UNA**
(Recovered from Crashed Alien Spaceship)

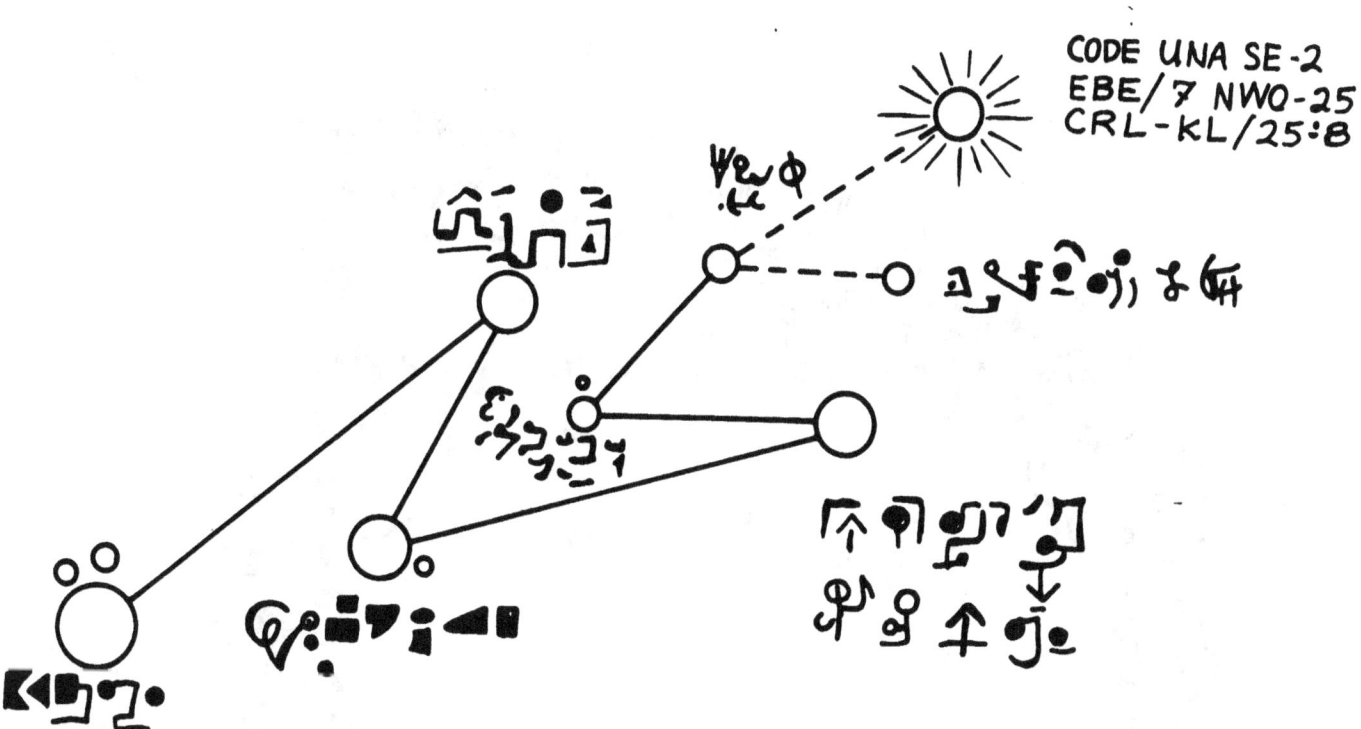

**ULTRA-TERRESTRIAL - CODE: ULTRON** (Energetic/Intelligent life forms without material body). This information was recovered from NCR/27B-01 MWC-2.B7 - AC "GODAR" PROJECT, as well as other crash sites.

This Alien is created in the centers of quasars. A type of sensitive energy living in the center of pure energy, and is a component more stable than the material Solioa that is found in our quasar.

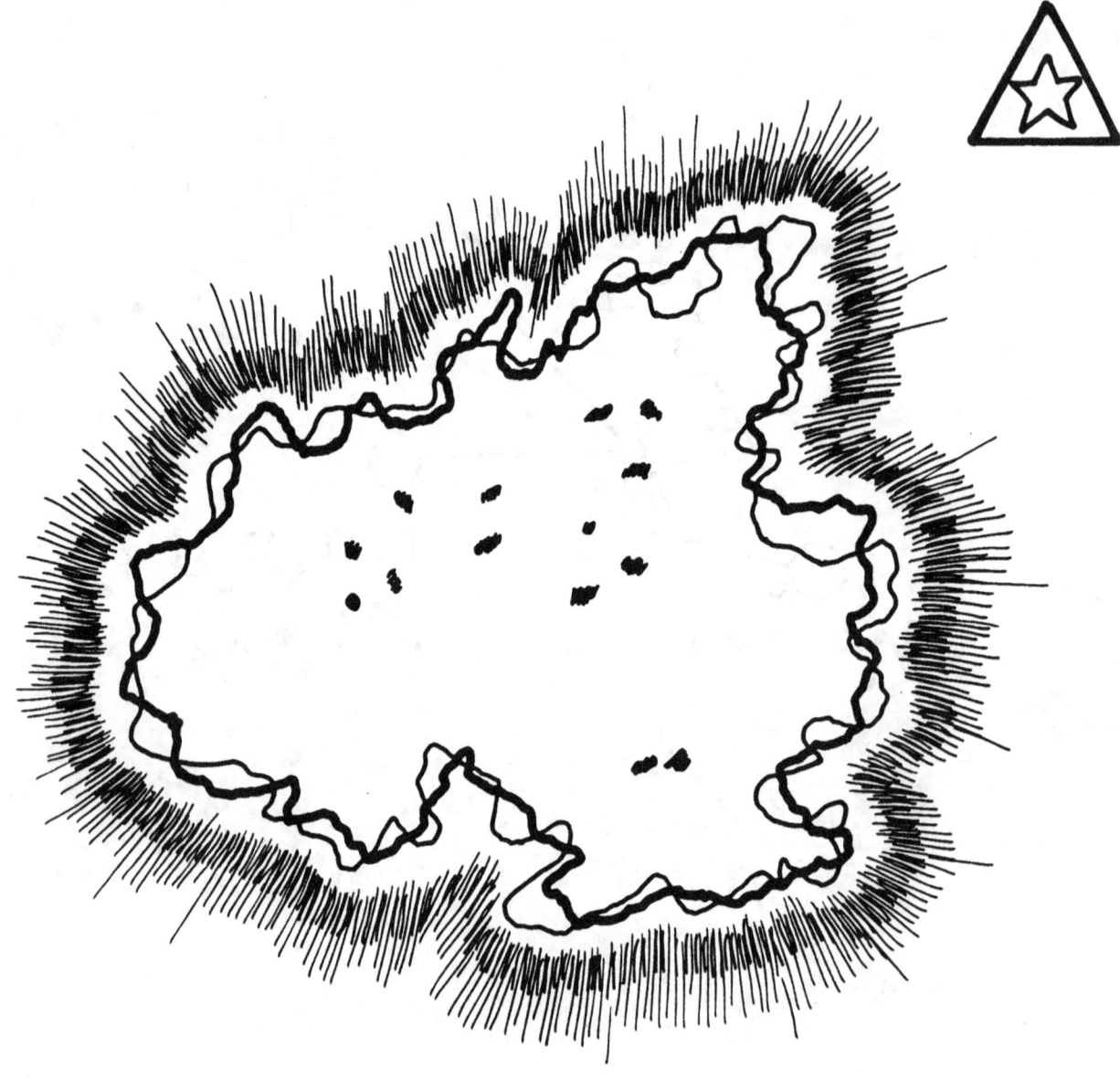

They evolved from a series of mono-electronics, to a series of more complex energy and created a collective civiliztion that does not know the individual thought or mind.

The Ultrons are an entity composed of primary mental energy, and radiates strong in the visible, ultraviolet and Kappa Radiation bands. As are "Nodes or Foir" serve as the origin of the Ultrons' radiant emissions. As the spectrograph proceeds through the Ultraviolet range from 103.2 to 112 angstrom units, we observe that these Foir are not constant for different wavelengths; by varying their relative positions, the Ultrons can control the intensity of their emissions. Ultrons perceive dimension and objects in terms of electromagnetic energy - knowing the speed of light through various mediums, it is able to judge its surroundings much as the Terran Bat is able to use the reflection of its own sounds to avoid flying into an object.

The radioactive nature of the Ultrons a non-corporeal being, best suited by the Hutchinson Radiation Analysis (HRA), (a graphic representation of any anergy source, in this case the life aura of the Ultrons are strongest in the ultraviolet range, the HRA is a standard readout on most sensor-equipped instruments appearing as a fluctuating moire' pattern.)

**INTRA-TERRESTRIAL - CODE: INTERAV** (People from outside Earth/Dimension/Time-Space of life at Planet Earth from ancestral time). This information was recovered from INAC/26.7B - AC 43.2, INAC-02, as well as other crash sites. They came to Earth several EON's ago and now Terra (or Earth) is their planet too.

We talk here about the seventh largest race, we found two large races living below the four earth capades....

### THE REPTLOIDS

Kingdom: Animal
Phylum: Vertebrata
Class: Reptilia
Order: Theropoua
Family: Hominidae
Genus/Specie: Homo Lacerate

LEVEL OF SPECIE: BETHA RDN-3
OBS: Survive only close to specific devices.

They live at a former society and have a high degree of technology.

### THE REPTLOIDS DATA

| | | |
|---|---|---|
| Average Height: | Male: | 2.0 Meters |
| | Female: | 1.4 Meters |
| Average Weight: | Male: | 200 Kilos |
| | Female: | 100 Kilos |
| Body Temperature: | Male: | Ambient Temperature |
| | Female: | Ambient Temperature |
| Pulse/Respiration: | Male: | 40/10 |
| | Female: | 40/10 |
| Blood Pressure: | Male: | 80/50 |
| | Female: | 80/50 |
| Life Expectancy: | Male: | 60 Earth Years |
| | Female: | 23 Earth Years |

*f*

**REPTILOID**
Type A
Male

**45**
45

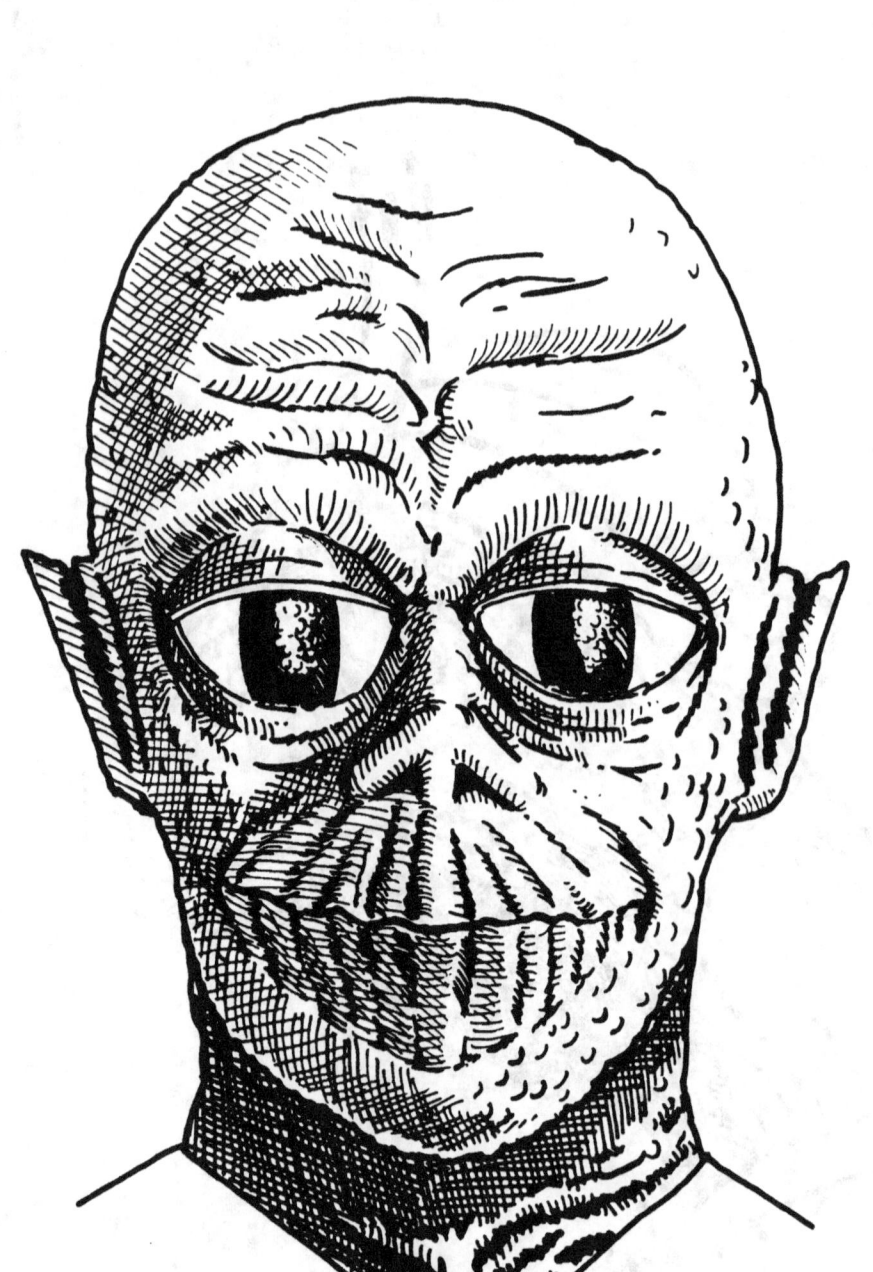

**REPTILOID**
Type B
Male

Cold-blooded like all reptiles, the Reptiloid is found to flourish in a warm, tropical clime (normally artificial at the big caves). With imperfect respiration providing just enough oxygen to supply tissues and maintain the processing of food and combustion, their temperature can be raised only a few degrees above the ambient. The reproductive system is ovouniparous, with eggs hatching in the oviduct prior to birth. The underdeveloped Reptiloid (for faster activities, physical activities) cerebellus results in a slowness and simpler city of movement. The Reptiloid eye is composed of thousands of microscopic facets, each facet with its own independent protective lid. The eye is almost never closed entirely during waking hours; rather, section of the organ is shut down in conjunction with the dominant light source.

**NOTE:** The Reptiloids or Reptoids survived "Hidden" inside the Earth at a Big Cave Underground surface.

Photograph

**REPTILOID OR REPTOID**
**Type C**
**MALE**

**News Clippings**

***Los Angeles Times*, Thursday, July 21, 1988..... *Newsmakers*.....** It's a Hulk - Christopher Davis 17, swears the creature that attacked him while he was changing a tire in the middle of the night in Scape Ore Swamp near Browntown, S.C., was 7 feet tall and had red eyes and three fingers on each hand. Then Tom and Mary Waye reported that their car had been "chewed up" at the same location. Was it the Lizard Man, as the locals are calling him, or just a red fox or a "muddy drunk" -- the theory of state biologist Matt Knox, who has been called in to help investigate the sightings. Sheriff Liston Truesdale said he has been swamped with phone calls from people claiming to have seen the slimy critter, "and these are reputable people". The run with television crews and observers "hoping to catch a glimpse of Lizard Man, and a Columbia News station is offering a $1-million reward for his capture.

***Los Angeles Herald Examiner*, Thursday, July 21, 1988..... *It's Lizard Man I*.....** Story of 7-foot 'Lizard Man' puts the creeps into town..... BROWNTOWN, S.C. -- The sheriff has been hearing a lot about the "Lizard Man", 7 feet tall with red eyes and three fingers on each heand, but a state biologist says everybody would be better off looking for a red fox or a muddy drunk.

Christopher Davis, 17, told Sheriff Liston Truesdale he had been attacked several weeks ago by the creature in Scape Ore Swamp as the teen-ager was changing a flat tire about 2 a.m.

Truesdale said he is getting other calls from people who said they saw the creature, "and they are reputable people" -- Gary C. Fong, from news service reports.

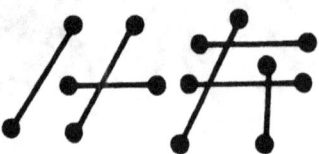

**SERV 58.6 ABR - CB-7 PANT. BOLHN 2E - AC (MULDOSV)**

## THE INSECTOIDS

Kingdom: Animal
Phylum: Vertebrata
Class: Mammalia
Order: Primata
Family: Insectoid
Genus/Specie: Homo Insectoid

LEVEL OF SPECIE: Unknown
OBS: Unknown

## THE INSECTOID DATA

| | | |
|---|---|---|
| Average Height: (Master Race) | Male: | 1.6 Meters |
| | Female: | 1.2 Meters |
| Average Height: (Servant Race) | Male: | 1.0 Meters |
| | Female: | 1.0 Meters |
| Average Weight: (Master Race) | Male: | 70 Kilos |
| | Female: | 40 Kilos |
| Average Weight: (Servant Race) | Male: | 35 Kilos |
| | Female: | 35 Kilos |
| Body Temperature: | Male: | 102 degrees Fahrenheit |
| | Female: | 102 degrees Fahrenheit |
| Pulse/Respiration: | Male: | 110/2 |
| | Female: | 110/2 |
| Life Expectancy: | Male: | 130 Earth Years |
| | Female: | 150 Earth Years |

**INSECTOID (Masters)**
Type 1
Male

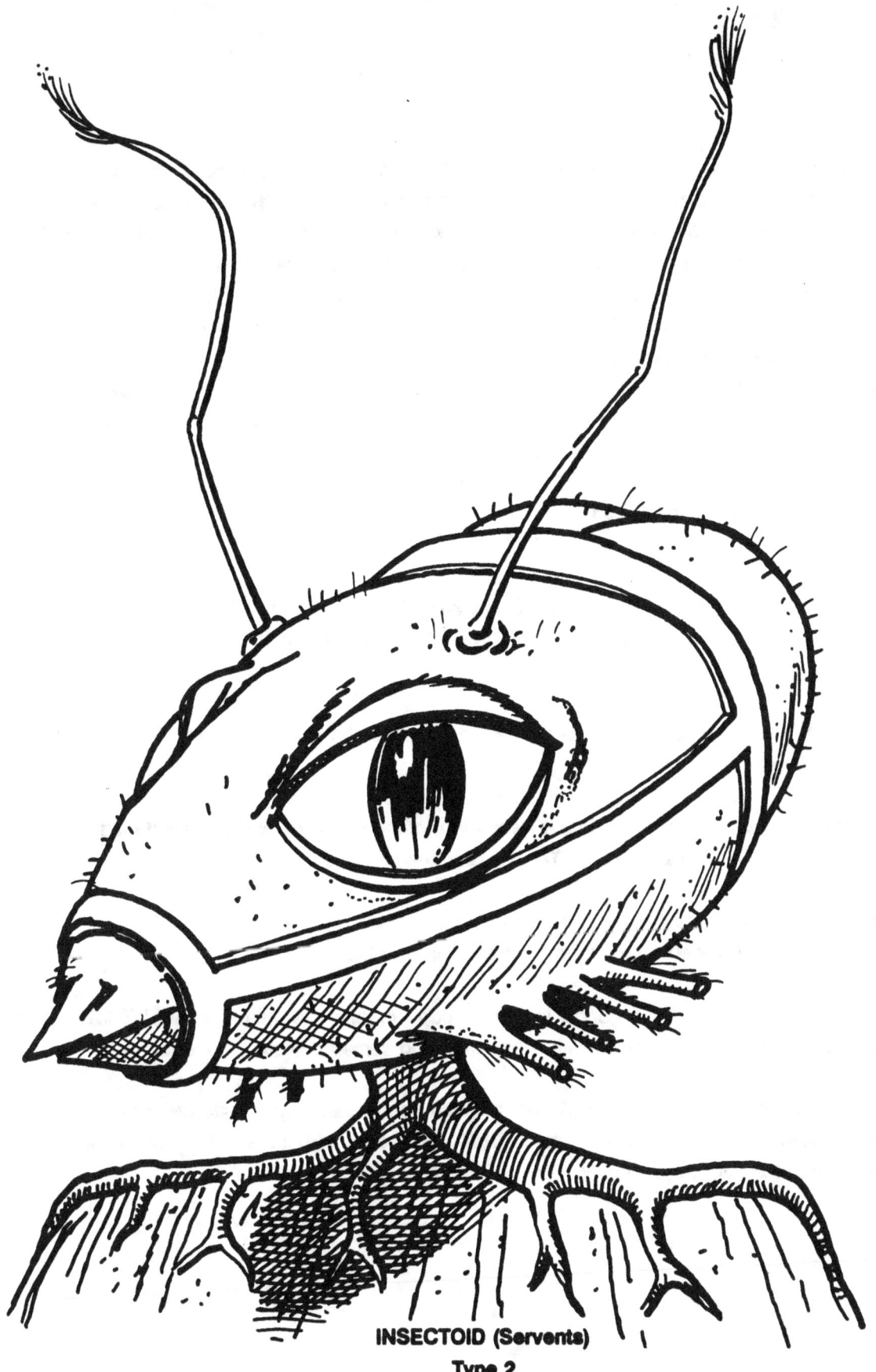

**INSECTOID (Servents)**
**Type 2**
**Male**

The Insectoid retina is composed entirely of tone-sensitive rods, and is incapable of discrimination between different wavelengths of light. Therefore, the addition of color to the insectoids vision is accomplished by the dual antennae which, in addition to being auditory receptors, are made up of a complex network of wavelength-sensitive cones. Owing to the highly directional nature of the antenna, the corona of vision is perceived by the subject in tones of grey. Because of this correlation of four independent light-receptive organs, Insectoid vision can be correctly termed "Quadroscopic", resulting in relatively (Humanoid) superior depth perception. Insectoid auditory capacities are highly developed, and Insectoids are capable of distinguishing from among a wider range of audio frequencies than is normal for humanoids. Because of the mono-directional antennae, Insectoids usually listen with their head tilted slightly downward.

NOTE: They have a limited exoskeleton.


**SERV 59.60, 61, 62, 63 ABR - CB-7 PANT. BOLHN 28 - AC (MULDOSV/PLUS)**
Several races, including some already discussed, make their home in the Inner-Earth. These are as follows:

- **Green People** - Part of the survivors, survived as part of the ancient K'ushul K'hotans or green people.

- **Caspan Jassan Paegans or Golden Race** - Part of the survivors, survived inside big group of mountains at Orient part of Earth. They created the **Essenis**.

- **The Eternals** - Some surviving people from Altair genetic mutated, which caused getting older to be incredibly slowed down, enabling them to live for EON's.

- **The Kasimar or Blue Race** - Some surviving people created the people of MU, LEMURIA and ATLANTIS. The survivors now live in inner-Earth and some Atlanteans live underwater.

- **The Deviants** - Are surviving people that mutated from Kasimar/Insectoids (**very dangerous and primitive creatures**). The legend of DEMONs was born from the underground Deviant race. See the next page for sketch of the Deviant.

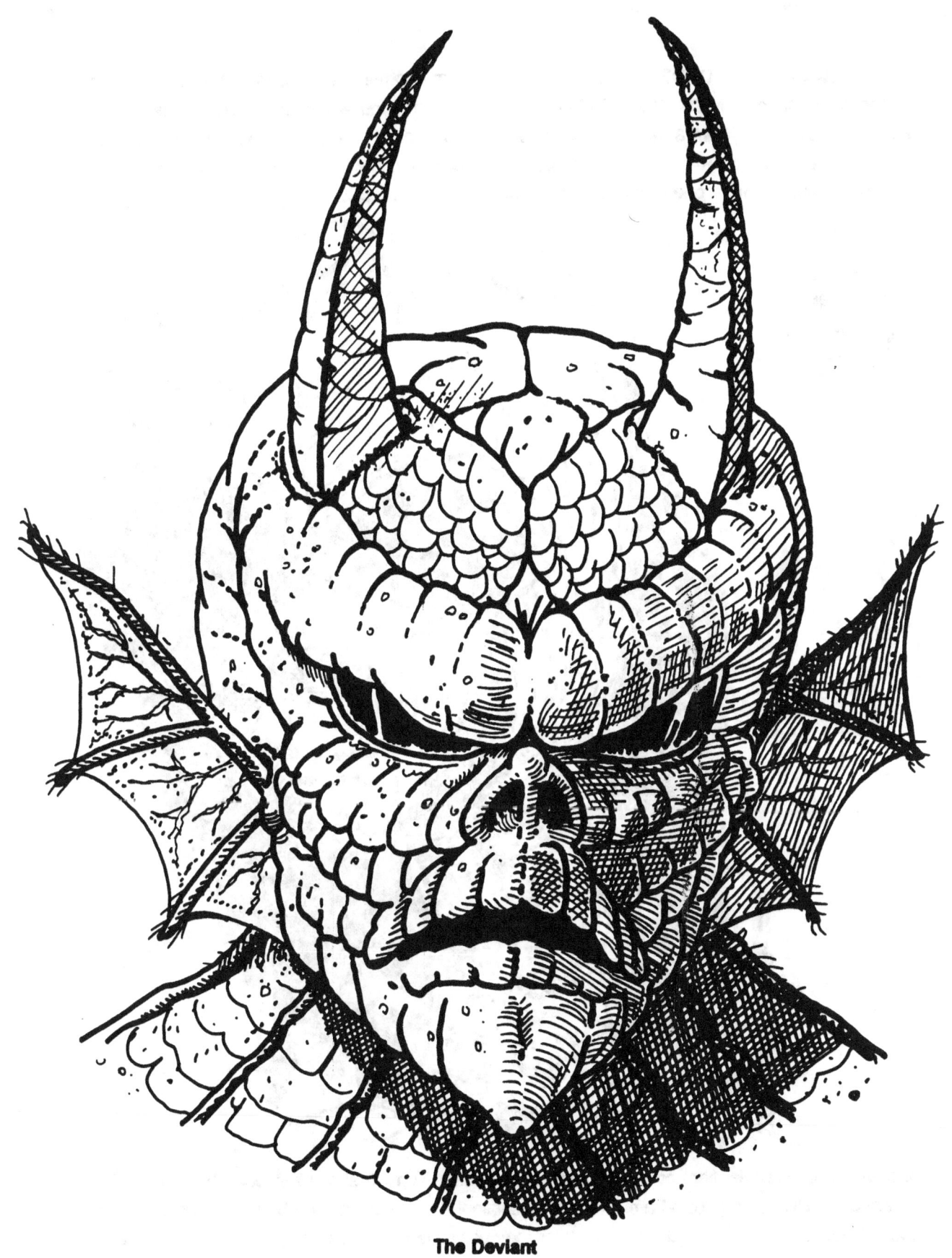

**The Deviant**

**META-TERRESTRIAL - CODE: MEREDITH** - (Aliens from another Time/Space or Earth people from another Time/Space). META-TERRESTRIAL visitors are from another time and space. This information was recovered from **CRM/26.06 - ABR-26**, as well as other crash sites. **Time Travelers** - Our files take only the Essessani Race. A cross-breeding between Humans from Earth and Extra Terrestrial from Zeta-Reticulae from one of our possible futures.

Of course it is possible there could be other races of Meta-Terrestrial, we just don't have enough data yet about it. **NOTE:** You found a example of Essessani Race watching the Steven Spielberg's movie "Close Encounters of the Third Kind" (Look for the special version, it's better).

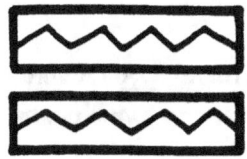

**PARA-TERRESTRIAL - CODE: PARAMUS** (Aliens from multi-universe). The Para-Terrestrial come from several different parallel universes from the Multiuniverse. When they accelerate or decelerate the frequency of vibration of their atomic structure they are able to exist at our 3rd dimension. This information was recovered from IACT/26, OBR-32 - AC 204 (BANTY), as well as other crash sites.

We can't tell much about these beings, because we don't have enough Data about it yet.

1. The name of this creature is almost impossible to repeat or even to write. We call this intelligent creature **XZ**. They look like one single Ameba or Jelly.

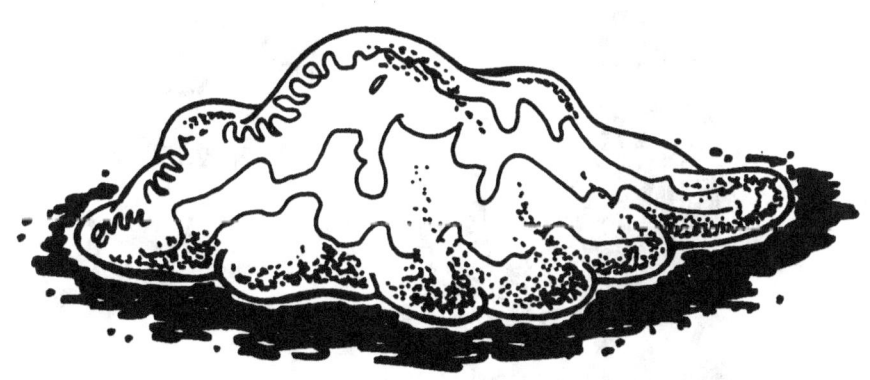

**The XZ Creature**

2. The **MOOK** - A very small (1.3 Meter) and hungry creature with long arms and three fingers, wearing funny cloths.

The MOOK

3. The **WADIG** - Looks like a primitive kind of primate but he is an intelligent being. Normally he never wears clothes (this fact, makes strong the idea some people believe that the **WADIG** is some kind of primitive missing link)

We can continue talking about Para-Terrestrial but is not logical because we don't have necessary quantity of Data for making good research. This is because we talked only about the **X2**, The **MOOK** and the **WADIG** just for literal illustration of matter.

# GREYS

**EXTRA-TERRESTRIAL - CODE: EBEs** (Aliens from 3rd dimension, or from outside earth). This information was recovered from IAC - XXXX - P14AB and IAC - XXXX - P14NR, as well as other crash sites.

**Extra-Terrestrial**

Wrongly designated ALFs, (Alien Life Forms). I said that because all other five types are ALFs also. All six types of aliens are ALFs.

We have around 160 different species at our biosphere from several points in the universe. They are Humanoids, Human and Not Humans.

We talk now about the other five Alien races that stayed much longer than any other race and have much more influence during our history.

1. Type A:      Rigelian, Rigel, or Gray
2. Type B:      2-Reticulae, Z-Reticulae 1, or Grey
3. Type C:      2-Reticulae, Z-Reticulae 2, or Grey
4. Type D:      Orion, Pleiades, or Nordic
5. Type E:      Barnard Star, Orange

**THE FIRST EBE - IAC - XXXX - P14SC**

EBE is the name or designation given to the live extraterrestrial Alien captured at the 1947 Roswell, New Mexico crash. He died in captivity.

**NOTE:** O. H. Cril was a "Joke" created by John Lear and other friend for personal purposes. No body knows why Lear denies the KRLL existence.

**KRLL**

KRLL or KRLLL or CRLL or CRLLL pronounced Crill or Krill was the hostage left with us at the first Holloman landing as pledge that the Aliens would carry out their part of the basic agreement reached during that meeting. KRLL gave us the foundation of the **YELLOW BOOK** which was completed by the Alien guests at a later day. KRLL became sick and was nursed by Dr. G. Mendoza who became the expert on Alien biology (Exobiology). His information was disseminated under the pseudonym O. H. Cril or Crill. KRLL became the Regelian Ambassador to the United States. There have been rumors of KRLLs death. We don't know if they are true or not.

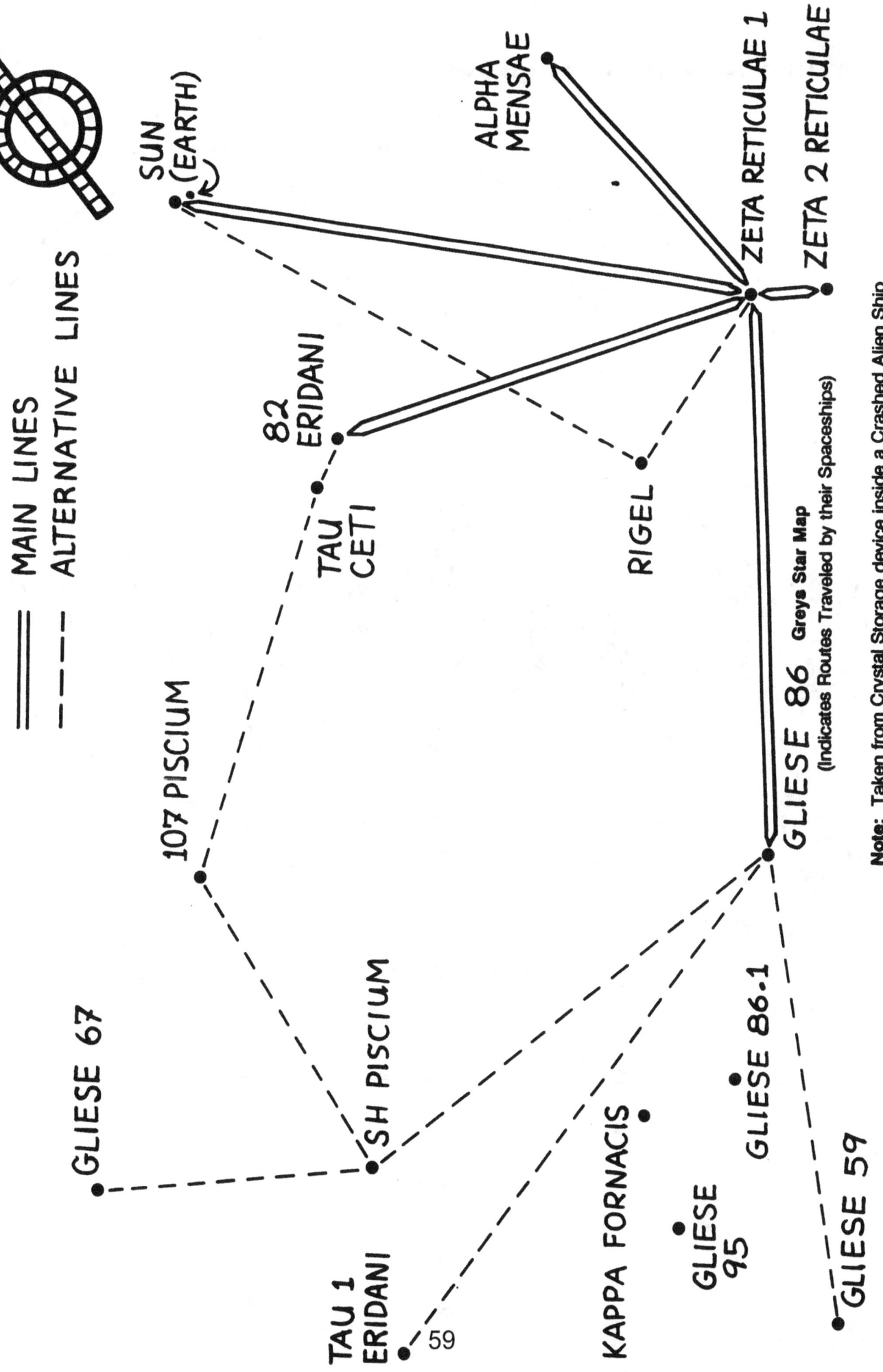

**RIGELIANS** - Also called Malevolent Alien Life Form (ALF).

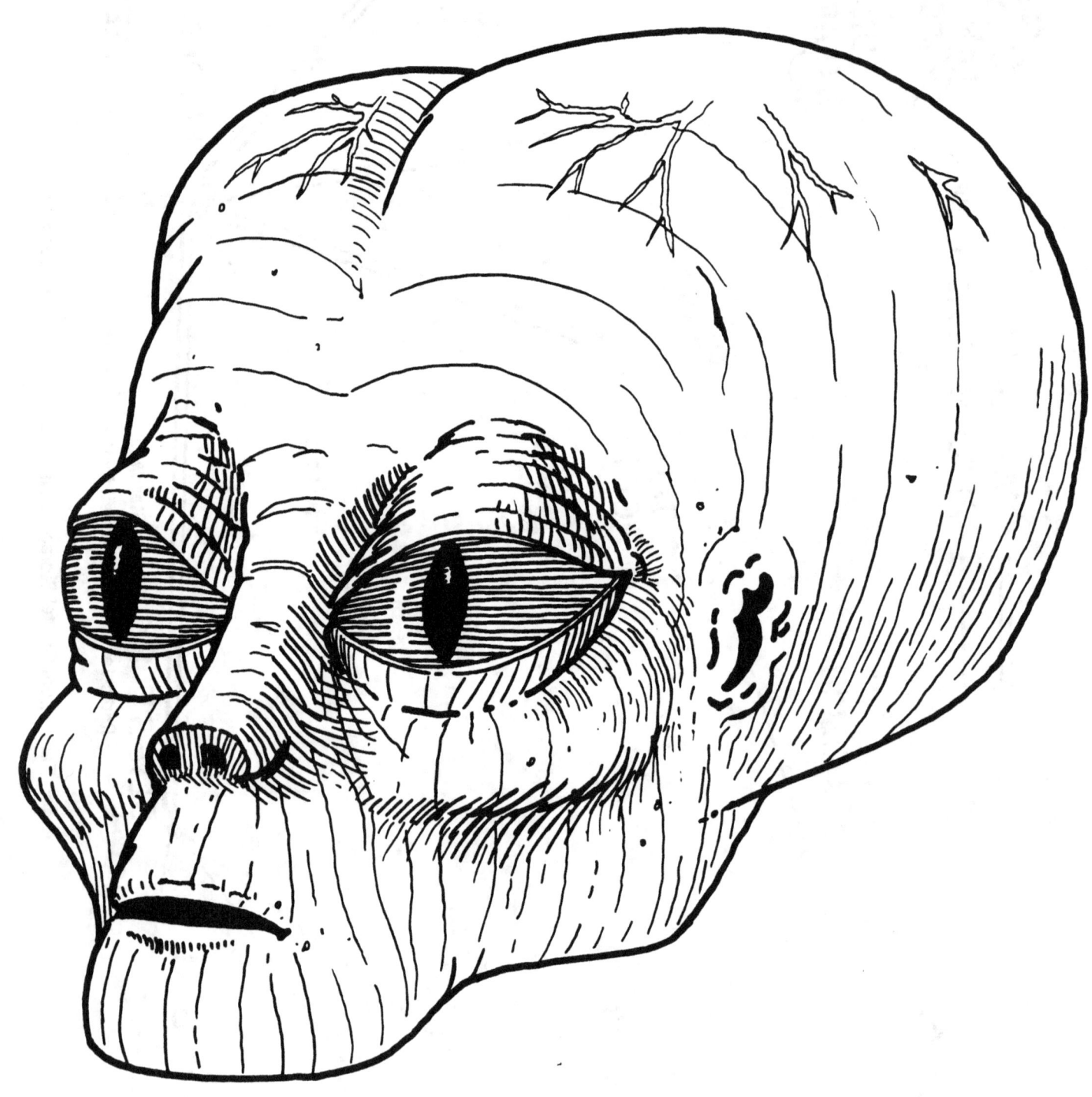

**REGELIAN**

The Typical Malevolent ALF as represented thus far can be described as follows:

- Between three (3) to five (5) feet in height.
- Erect standing biped. Long thin legs.
- Small build (thin).
- Head larger than normal, (to human proportions).
- Absence of auditory lobes, (external ear lobes).
- Absence of body hair.
- Large, tear shaped eyes, opaque black with verticle slit pupils, (cats eyes).
- Eyes slanted approximately 35 degrees.
- Small straight mouth, with thin lips.
- Arms resemble praying mantis, (normal attitude), arms reach to knee when extended.
- Long hands, with small palms.
- Claw like fingers, (two short, two long webbed fingers).
- Tough, grey skin, reptile like in texture.
- Small feet with four (4) small claw like toes.
- Some organs are similar to humans but developed in a different evolutionary process.
- The most significant finding is that they have a nonfunctioning digestive system and two separate brains. The digestive system in those examined were atrophied, conforming to absence of provisions in recovered crash. This however, is still just preliminary findings.
- Movement is deliberate, slow, precise.

- Secondary findings, after crash site study - Alien subsistence requires that they must have human blood and other human biological substances to survive. In extreme cases (circumstances) they can subsist on other animal fluids. Food is converted to energy by Chlorophyll through photosynthesis and waste products are excreted through the skin, (some kind of botanical life form ?). These creatures possess two separate brains separated by the mid cranial lateral bone partition, (anterior brain, posterior brain) with no apparent connection between the two.

* **ALFs** furnished extensive information on the Aliens and their history which is known as the **"YELLOW BOOK"**.

* Because the Aliens have a tendency to LIE, we can't be 100% sure about all the information inside the **"YELLOW BOOK"**.

* **LUNA-1** is the Rigelian base at the other side of the Moon, or the far side of the Moon like some scientists prefer to say. It was seen and filmed by the Apollo Astronauts. A base, a mining operation using very large machines, and the very large Alien Crafts described in sighting reports as "Mother Ships" exist there.

* **WAVENEST** is the Regelian base in the Atlantic Ocean. A under water Alien base, mining and big

Cigar shaped crafts.

**The Rigelian Saga** - The original contact between the government and the Extra-Terrestrial Biological Entities (EBEs), who normally are Grey in color and about three and a half to four and a half feet high and coming from the Rigel Star System (hereafter referred to as the Greys) was achieved between 1947 and 1971. We knew that the Greys were instrumental in performing the mutilation of animals, as well as some humans and that they were using the glandular substances derived from these materials for food (absorbed through their skin) and to clone more Greys in their underground laboratories. The government insisted that the Greys provide them with a list that would be presented to the National Security Council (this list has never been completed). Through all this, the government thought that the Greys were basically tolerable creatures, although a bit distasteful. They presumed at the time that it was not unreasonable to assume that the public would and could get used to their presence. Between 1968 and 1969 a plan was formulated to make the public aware of their existence over the suceeding twenty (20) years. This time period would culminate with a series of documentaries that would explain the history and intentions of the Greys.

The Greys assured the government that the real purpose of the abductions was for monitoring of our civilization, and when they learned that the abductions were a lot more frequent and insidious than they were led to believe, the government became concerned. Their concern was also based on additional information regarding the purposes for abductions. By the time the government had found out the truth about the intentions of the Greys, they, the Greys, intended to stay here, where they always will stay in control of Earth, "it was too late". The government by mistake had already "Sold Out" humanity. Not that it would have made any difference, because they were here doing what they were doing anyway.

In 1983, a story was outlined by the government sources that said that the Greys are responsible for our biological evolution through manipulation of the DNA of already evolving primates on this planet, they don't like to tell anybody that the real Rigelians are Nordics and they found survivors from one big catastrophe, the ancestral seven races, the Uni-Terrestrials. Various time intervals of the DNA manipulation were specified for 25,000, 15,000, 5,000 and 2,500 years ago originally, the government thought that the Greys meant us no harm, but in 1982 to 1988 the picture that emerged was exactly the opposite. The story now is one great deception at several different levels. The Greys "Trojan Horse" style manipulation and lying, which involved MJ-12/Majic forces, from over four decades ago, now it's 1990, the government still uses disinformation on the subject of UFOs, in order to perpetrate the agreement with the Greys, free of public scrutiny; the lies to the abductees, the Greys ongoing abduction of people and mutilation of animals in order to harvest enzymes, blood, and other tissue for their own survival needs, and a genetic blend of the Grey race and a tall Nordic act to enable Grey interface with humans to be done with greater ease. The apparent reasoning for the Grey preoccupation with this is due to their lack of a formal digestive tract and the fact that they absorb nutrients and excrete waste directly through the skin. The substances that they acquire are mixed with hydrogen peroxide and painted on their skin, allowing absorption of the required nutrients. It is construed from this that some weaponry against them might be geared in this direction.

The innercore of the CIA is deeply controlled by the Greys. The CIA view's interaction with the Greys as a path to greater scientific achievement. One reason for so many UFOs is that other cultures are watching with extreme interest. Scientists from other cultures arrive to watch. The Greys also are working with some groups of UFO fanatics, lying to them, and using them. The ultimate evil is masked and is of psychological complacency that leads one to adhere to a group philosophy rather than eke out one's own horizons. As soon as you acquire an awareness of being a so called "chosen special group" you are on the road to a fall. That is the seed of destruction in any society and any culture and it leaves it vulnerable. It will be the eventual undoing of the Greys as well. They see not their error, it is the very weakness they seize upon that is their own inherent weakness. To try and change a Grey, or a mistaken cult type of "Star Person" (following the Grey plan) is futile. It will happen but all in its own good time, it is the spirit that makes anyone stand up and disagree with something that is untrue and incorrect that will be the thorn in the side of the Greys, and other forces that have allied with them.

**Alien Bases at SOL III Perimeter** - During the occupation of the Greys, they have established quite a number of underground bases all over the would, especially in the United States. One such base, among other in the same state, is under Archuletta Mesa, which is about two and a half miles Northwest of Dulce, New Mexico. Details about that base come across by way of several different sources. This is a Kilometer underground base beneath *Archuletta Mesa* and the *Jicarilla Apache Indian Reservation* since 1976, one of the areas of the U.S. hardest-hit by mutilations. This installation is operated jointly as part of an ongoing program of cooperation between the U.S. government and EBEs. There are also underground bases at *Kirtland and Holloman Air Force Bases,* as well as, at scores of other bases around the world, including *Bentwaters, England*.

Back to the base under discussion, "Archuletta Mesa" we have the following data on that base. The base is two and a half miles Northwest of Dulce, and almost overlooks the town. There is a level highway thirty six (36) feet wide going into the area. It is a government road, and one can see telemetry trailers and five (5) sided buildings with a dome. Next to the dome, to the North there is a launch site. There are two wrecked ships there. One can see thirty six (36) feet long oxygen and hydrogen tanks near them. The ships that we got to, near the tanks, were atomic powered with plutonium pellets. Refueling of the plutonium is accomplished at Los Alamos.

The base has been there since 1948. The base is 4,000 feet long and helicopters are going in and out of there all the time. In 1979, something happened and the base was temporarily closed. There was an argument over weapons and our people were chased out. The Aliens killed sixty six (66) of our people and forty four (44) people got away. One of the people who got away was in fact a CIA agent who, before leaving, made some notes, got some photos and video tapes, and went into hiding. He has been in hiding ever since, and every six months he contacts each of five people he left copies of the material with. His instructions were that if he missed four successive contacts, the people would do whatever they want with the materials.

Somehow, a descripteion of the *Dulce Papers* were issued, and was received in December 1987 by many researchers. The Dulce Papers were comprised of some twenty five (25) black and white photos, a video tape with no dialogue and a set of papers that included technical information regarding the jointly occupied U.S. and Alien facility one (1) Kilometer beneath the Archuletta Mesa near Dulce, New Mexico. The facility still exists and is currently operational. It is believed that there are four (4) additional facilities of the same type, one being located a few miles to the Southeast of Groom Lake, Nevada.

A general description of what these papers contain is that they contain documents that discuss Cooper and Molybdenum, Magnesium and Potassium but mostly papers about Bill Copper. Sheets of paper with charts and strange diagrams. Papers that discuss Ultra Violet light and Gamma Rays. These papers tell what the Aliens seek, "The Secret of The Metagene" and how the blood, taken from cattle and some humans is used. They tell why some Aliens seem to absorb molecules to eat by putting their hands in blood, sort of like a sponge, for nourishment. It's not just food they want, the DNA in cattle and humans is being altered.

Why a "Type-One" creature is used as a Lab Animal, and how to change the atoms to create a temporary, almost human being. It is made with animal tissue and depends on a micro computer to simulate memory, a memory the computer has withdrawn from another human being, creating some kind of Clone. The almost human being or Clone is a little slow and clumsy. Real humans are used for training, to experiment with and to breed with these "almost humans". Some are kept in large tubes, and are kept alive in an *amber type liquid*.

Some humans are brain washed and used to distort the truth, certain male humans have a high sperm count and are kept alive for that reason. Their sperm is used to alter the DNA and create a non-gender being called "Type-Two". That sperm is grown in some way and altered again, put in wombs. They resemble ugly humans when growing but look normal when fully grown, which only takes a few months from Fetus size, but they have a short life span, less than a year. Some female humans are used for breeding, and subsequently, countless women have had a sudden miscarriage after about three (3) months into the pregnancy, and some never know they were pregnant, others remember contact some way. The fetus is used to mix the DNA in types one and two. The make-up in that fetus is taken at three (3) months and grown elsewhere.

That's what the "Dulce Papers" talk about, (some of this information is covered in volume two "The Pulsar Project"). There are some pen and ink reproductions of what one of the wombs look like i.e. two (2) by four (4) feet, an illustration showing one of the tubes in which the "almost humans" are grown, a page showing a simple diagram of crystalline metal, pure gold crystal, and what looks like either a genetic or metallurgical diagram or chart. Also attached is what looks like an X-Ray diffraction pattern and a diagram of hexagonal crystals, with a comment that they are best for electrical conduction. It would appear that the last half of the material in the papers applies to the super crystalline metal used for hull structure, or something along that line.

Obviously, this is all rather bizarre from a certain point of view, any point of view, in fact, material that is supported by years of descriptions and multitudes of corroborations must mean something, especially when compared against what seems to be going on. It is apparent from this, and other data that has been accumulated over the years, that there are underground bases and tunnel complexes are all over the world, and that more are being constructed all the time. Many of you may recall the "Shaver Mysteries" and inner-city stories, well, all that is true. There are cities down there, amongst other things, and some of them have nothing to do with the main subject of this document. They've been under there a long time.

Let's change direction for a moment, one individual by the name of Lew Teri has been working on some ideas regarding Geomagnetic Anomalies, which UFOs may use. We will go into what he has discovered, although the concept of the relationship is not new, and let you judge that for yourself.

After purchasing Aeromagnetic and Gravitational Anomaly Maps for the United States Geological Survey, it became evident that there was indeed a valid connection between these areas and UFOs. Mr. Teri gave a lecture in Arizona about that relationship, and was harassed by the FBI, and told that the information was and is sensitive. Mr. Teri took the hint and declined to talk publically about it to the degree that he had been doing. Both the aeromagnetic and gravitational (Bougier Gravity) maps indicate basic field strength, as well as areas of high and low field strength. Interestingly enough, the areas of maximum and minimum field strength have the following:

- All of these areas have frequent UFO sightings.
- All are either in Indian Reservations, Government land, or Government is trying to buy up the land.
- Many of them, especially where several are clustered together, are suspected base areas and/or areas where mutilations and abductions have historically taken place.

In these observations, Mr. Teri has gone far, but he has gone a little further in noting that there are times when UFOs are seen in these areas. Through pain staking research Mr. Teri found that the sightings, as well as, many abductions and mutilations occur:

- When there is a New Moon or within two (2) days before a New Moon.
- When there is a Full Moon or within two (2) days before a Full Moon.
- At the Perihelion (when the Moon is closest to Earth) or within two (2) days before the Perihelion.

A glance at the nearest Farmers Almanac will give you the information you require as far as the days for this year or any other time. There seems to be no concrete explanation for the coincidence of the times and the events, but appear to be true.

Back to our main subject of this chapter of this document, talking about Grey, Rigelian bases, we have the principal bases:

**United States of America (Earth)**
- Fairbanks, Alaska
- Groom Lake, Nevada
- Dulce, New Mexico
- Utah
- Colorado
- California
- Texas
- Florida
- Georgia
- Maine
- New York
- New Jersey
- Wyoming

**The Atlantic Ocean (Earth)**
- Wavenest

**The Moon**
- Luna-1
- Luna-2

**Obs-1** - No report for bases at Europe, Asia, Oceanic, Africa, Central and South America.
**Obs-2** - No information about Nordic bases or any other Aliens.

**Project BETA, The Study of Grey Psychology** - The Rigelians, either through evolvement or because the humanoid types are constructed, will exhibit tendencies for bad logic. They appear to have more frailties and psycho weaknesses than the normal humans. The Rigelians are not to be trusted, because of the Rigelians apparent logic system, a key decision cannot be made without higher clearance. All are under control of what they call "THE KEEPER", yet it would appear that even this is not the final authority. Delays as long as twelve to fifteen hours can occur for a decision.

Because of this apparent control, individual instantaneous decision making by the Rigelians is limited. If their plan goes even slightly out of balance or context, they become confused. Psychologically their morale is near disintegration. There is a pronounced dissension in the ranks, even with the humanoids, because of their own internal vulnerability, (mind-wise) to each other, and additionally their is a basic lack of trust between them.

They appear to be totally death oriented, and because of this, absolutely death fear oriented, ( this is a psychological advantage to us). The prime and weak areas discovered, probed, and tested are exactly what we had thought, their mind, "being their key strength", and that which they use to manipulate and control our minds. Manipulated in reverse, by using reverse psychology we could make them face a situation where they would have a vulnerable integrated weakness, we could use against them.

**Taxonomy of some Extra-Terrestrial Humanoids** - Working under the instructions of the humanoids for Rigel (Greys, type ones (1)), CIA and former Nazi scientists have developed and deployed malignant strains of bacteria and viruses, including AIDS, in order to exterminate undesirable elements of the human population.

The Greys, especially the type ones (1) are almost entirely devoid of emotions, but can obtain a "high" by "telepathically" tuning in the different kinds of intense human emotions, such as ecstasy or agony. Does that explain why UFOs have always been seen in regions of war or devastation, where humans are in conflict ?

We have in the United States several offspring of Alien and humans living together, this is part of a sociological experiment by our government.

Throughout recorded history, as well as during pre-historic times, there has been constant genetic manipulation of inter-breeding with Uni-Terrestrial survivors trying to save his or her cosmic hereditarily, in order to breed out the less evolved affected survivors. The Nordic races have participated in this form of new beginning, which makes us much more a part of them than we might suppose.

Greys have the ability to camouflage themselves as tall blondes Nordics through mental energy projection. Blondes Nordics never project themselves as Greys. Some Blondes seen with the Greys are physically real, but are prisoners of the Greys, who have either paralyzed them or have destroyed (neutralized) their ability to teleport through time and other dimensions.

Both blondes Nordics and Greys have the ability to disintegrate matter into energy and then reintegrate the energy back into matter. This ability allows them to pass through walls and to transport abductees out of their cars with the doors still locked.

The original Rigelians were the Nordics until they were invaded by the Greys, a parasitic race, who took over and interbred with them against their will. Several blonde Nordics escaped from the Grey takeover of their system and came to Earth because the original Rigelian and Uni-Terrestrials were the Nordics and claim to be the ones who seeded the Earth. It is because of this common ancestry that terrestrial humanity is of such interest to both the Nordics and the Greys. Greys were and are easily able to impregnate terrestrial human females either on board their ship or while they sleep in their homes. Males need not be manifested in visible form for this to occur.

The conflict between the Nordics and the Greys is in a state of temporary truce, although the conflict between Rigelians and the Syrius Star System is being fought actively. The Nordics with speech abilities will respond violently if attacked or threatened, but the telepathic ones will respond peacefully. Nordics were sometimes mistaken for Angels in earlier centuries. They do not seem to age, and consistently appear to be from twenty five (25) to thirty five (35) years of age. The Nordics now inhabit the Procyon Star System.

## GREYS
### TYPE A
Type A's are the large nosed Greys from the Rigel Star. These Greys are ones with whom our Government has a treaty.

### DATA
- The approximate height is 3.5 to 4.5 feet tall. Some approximately 5 feet. Their weight is approximately 40 pounds.
- Two semi-rounded black eyes, without pupils. They are large, almond-shaped, elongated, sunken of deep set, far apart, slightly slanted, appearing "Oriental" or "Mongoloid".
- The head, by human standards, is large when compared with the size of torso and limbs.
- No ear lobes or protrusive flesh extending beyond apertures on each side of the head.
- Nose is long and vague. Two nasal openings are indicated with only slight protuberanler.
- Mouth is indicated as a small "slit" without lips, opening into a small cavity-mouth, which appears not to function as means of communication or an orifice for food ingestion.
- Neck described as being thin; and in some instances, not being visible because of garment on that section of the body.
- The head is hairless.
- The torso is small and thin.
- Normally they wear a metallic but flexible garment.
- Arms are long and thin and reaching down to the knee section of the body.
- The hands have four fingers each, but no thumb. Two fingers appear longer than the others on each hand. They have some kind of long fingernails. A slight webbing effect between each finger exists.
- Legs are short and thin. Their feet are like an orangutans feet.
- They have no teeth.
- Their skin description is some kind of beige, tan brown, or tannish or pinkish grey color. The texture as scaly or reptilian, and as stretchable, elastic or mobile over smooth muscle or skeletal tissue. No striated muscle. Light perspiration and very particular body odor. Under magnification the tissue of skin structure appears to be a mesh, or like a grid network of horizontal perpendicular lines. They suggest that the texture may be similar to that of the granular-skinned lizards, such as the Iguana and/or Chameleon, and may be similar to at least on other type of alien humanoid.

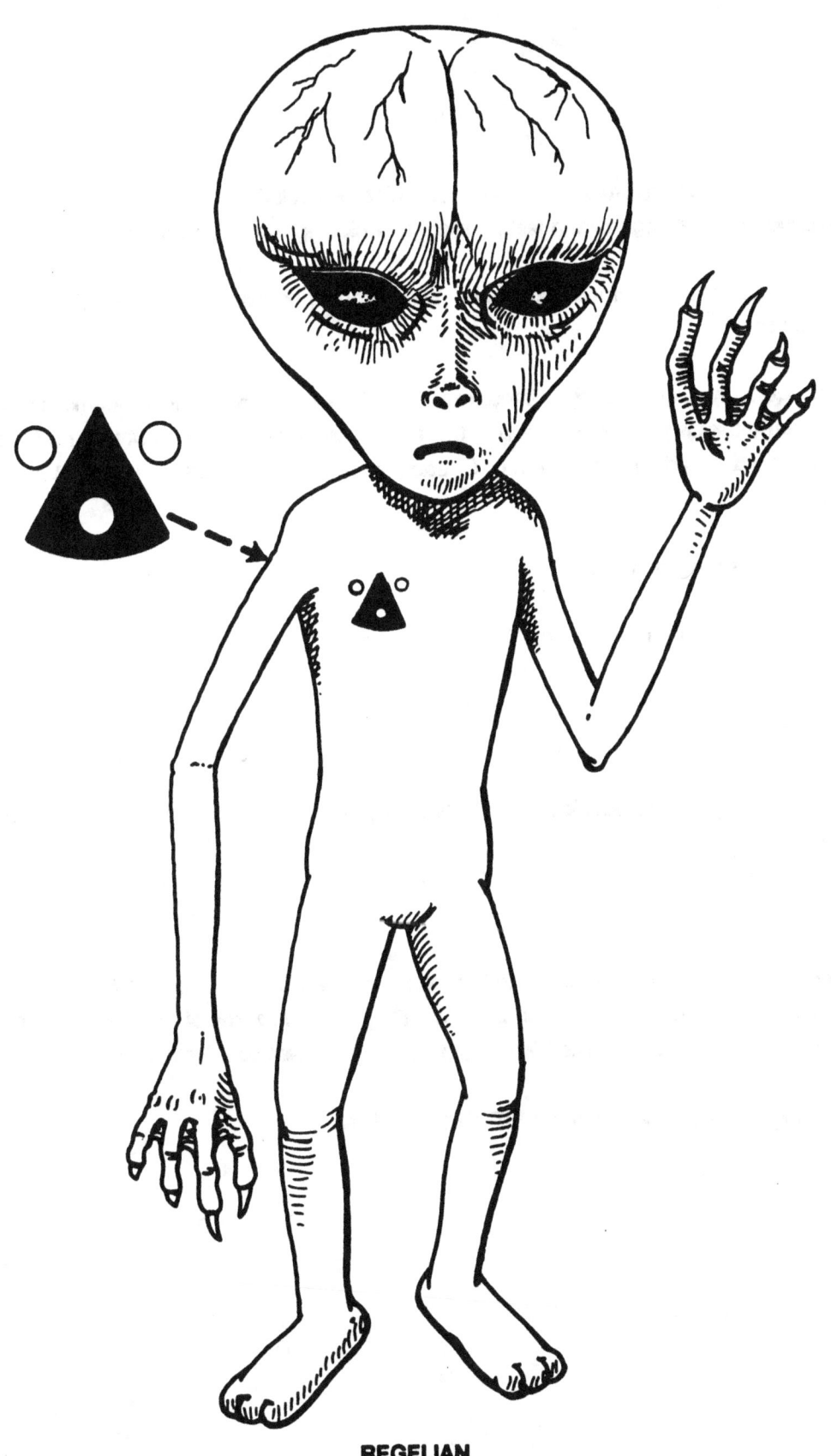

**REGELIAN**
Type A

- No apparent reproductive organs. Perhaps atrophied by evolutionary degeneration. No Genitalia. The absence of sexual organs suggests a system of cloning reproduction may be prevalent.

- The Aliens appear to be "formed out of some sort of mold", or sharing identical facial characteristics.

- Colorless liquid prevalent in body, without red cells. No Lymphocytes. Not a carrier of oxygen. No food or water intake is evident or known at this time. It is believed at present that no food may ever be required. No digestive system or GI tract. No intestinal or elementary canal or Rectal area described.

- Many variations of anatomy exist.

- Their life span is unknown at this time.

### TYPE B
Type B's are the Greys from Z-Recticulae 1. These Greys are also ones with whom our Government has a treaty.

### DATA
- Physiology more or less is very similar to that of the Rigelian Greys. They are Male and Female. The differences are they need food for survival. They prefer proteins of dairy meals. They have asexual form of reproduction. The "Orulo" grow outside the female Z-Reticulae 1.

The next page shows a sketch of a Z-RETICULAE 1 Alien.

**Z-RETICULAE 1**
Type B

## TYPE C

Type C, one and two, are the Greys from Z-Recticulae 2. These Greys are Alien researchers of us and our planet.

**Z-RETICULAE 2**
Type C

## DATA # 1

- Physiology is like Z-Reticulae 1's. The differences are at the eyes, they have big black pupils.
- They have ear lobes.
- The Type C number one Greys, have the same type of reproduction as the Z-Reticulae 1's.
- The mouth has small lips and a throat which they can speak from with verbal words.

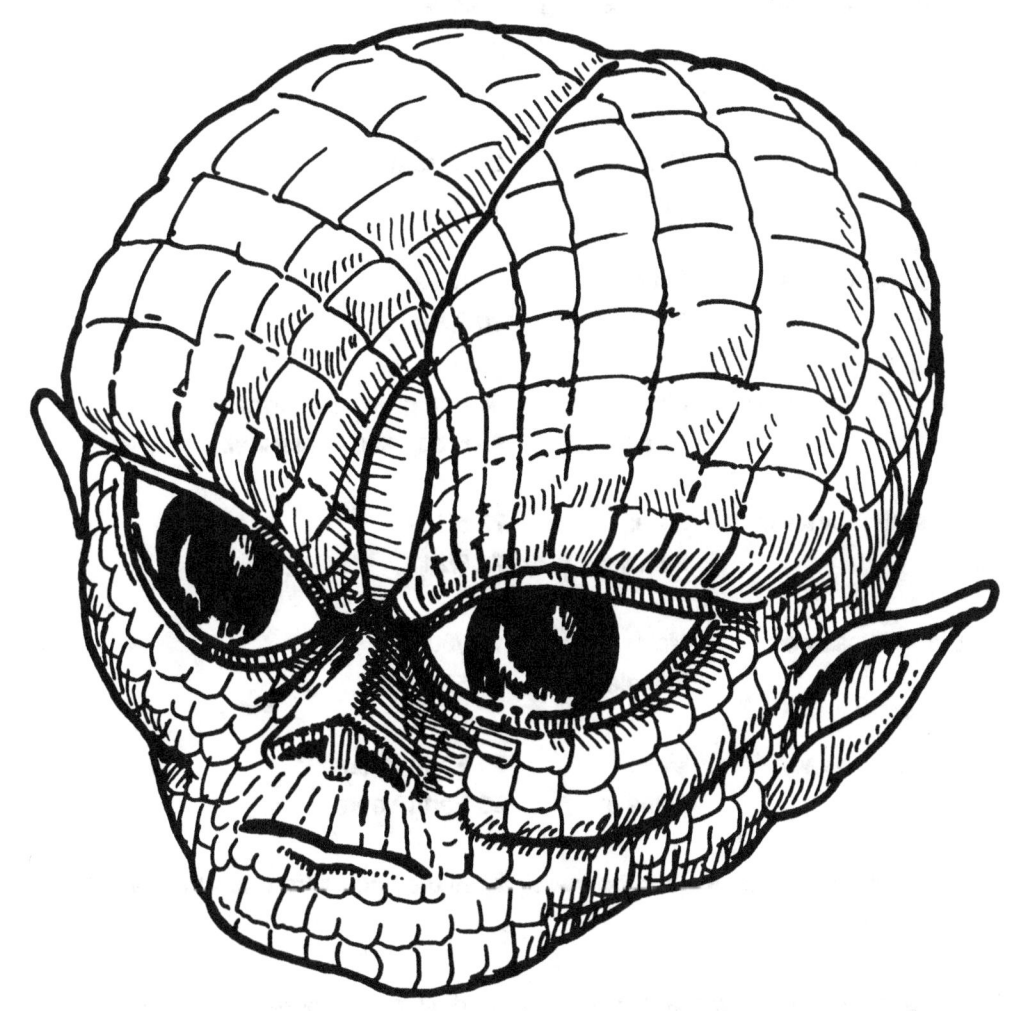

**Z-RETICULAE 2**
Type C

**DATA # 2**
- The differences are the shape of the head, see the sketch above.
- They eyes and eyeball.
- The Type C, number two Greys, have pointed ear lobes.
- They are smaller then 1st type of Z-Reticulae 2's.
- Their reproduction is accomplished by an egg.

# NORDICS

## TYPE D

Type D's are the Nordics. The Nordics are normally a blond humanoid type.

## DATA

| | | |
|---|---|---|
| Average Height: | Male: | 2.0 Meters |
| | Female: | 1.7 Meters |
| Average Weight: | Male: | 90 Kilos |
| | Female: | 70 Kilos |
| Body Temperature: | Male: | 98.6 degrees Fahrenheit |
| | Female: | 98.6 degrees Fahrenheit |
| Pulse/Respiration: | Male: | 72.5/16 |
| | Female: | 72.5/21 |
| Blood Pressure: | Male: | 120/80 |
| | Female: | 80/50 |
| Life Expectancy: | Male: | 60 Earth Years |
| | Female: | 23 Earth Years |

They come from several places, but the main places they come from are the Pleiades and Orion star system.

NOTE: I have found some smaller Nordics with dark skin or dark hair. There is still not enough data about this offshoot.

# ORANGES

## TYPE E

Type E's are the Oranges. The Oranges are normally a red haired humanoid type. They come from the Barnard Stars.

## DATA

|  |  |  |
|---|---|---|
| Average Height: | Male: | 1.7 Meters |
|  | Female: | 2.0 Meters |
| Average Weight: | Male: | 70 Kilos |
|  | Female: | 50 Kilos |
| Body Temperature: | Male: | 91 degrees Fahrenheit |
|  | Female: | 91 degrees Fahrenheit |
| Pulse/Respiration: | Male: | 242/61 |
|  | Female: | 242/61 |
| Blood Pressure: | Male: | 80/40 |
|  | Female: | 80/40 |
| Obs: |  | They can use a beard sometimes. |

# ALIEN TECHNOLOGY

**ALIEN CRAFTS WITH DIMENSIONAL FACTORS AND/OR DIMENSIONAL ORIGIN.** To understand this concept you have to realize we live in a Universe made of ten (10) Dimensions; nine (9) Spatial and one (1) Temporal. On Earth though, we just use four (4) Dimensions; three (3) Spatial and one (1) Temporal, because on Earth six (6) dimensions are compacted at one Abstract Dimension. Following the theories of two physicists Dr. John Schwartz, (USA) and Dr. Michael Green, (England) the Particles of our Atomic Structure have the consistency of ten (10) dimensions. For the system to be established, after the BIG BANG, the dimensions collapsed and created the compacted effect of the six (6) other dimensions into one dimension.

The Aliens used this knowledge, which they call the **NORDWAG FACTOR**, which increases one large field of acceleration or particles and recreates a new dimensional spaces. This means that the inside of the Alien craft is much bigger than the outside size to the Alien craft is in size and dimension.

Some similar aspect affect over 3rd Dimension (Space) and 4th Dimension (Time) to the six (6) other dimensions: MULTIVERSE (MULTI-UNIVERSE Theory) and not a single UNIVERSE Theory. Universe is a single part of the total of the MULTIVERSE.

**THE U.S. GOVERNMENT AND ALIEN TECHNOLOGY.** Of course they (Aliens) have given high technology to the Government but how do scientists, (like myself) know the difference between a toy or a real sophisticated device if they never have seen one before ? They give us primitive concepts in computer systems (micro-chips and integrated circuits) and we improve much more faster than they expect we can. After this, Aliens have been much more careful about what "Toy" they give us because we have demonstrated that we can turn a "Toy" into High Technology in short order.

The Main interests of the U.S. Government is to collect all possible Alien information in science (especially new sources of energy), technology (specially beam weaponry and tactical Aircraft), medical (extend the life expectancy of Humans and total elimination of deadly infirmities, i.e for elimination of "the weak" people from our society), and Intelligence (especially in mind control). All of these objectives are still ongoing.

## The Implant or S.B.M.C.D.

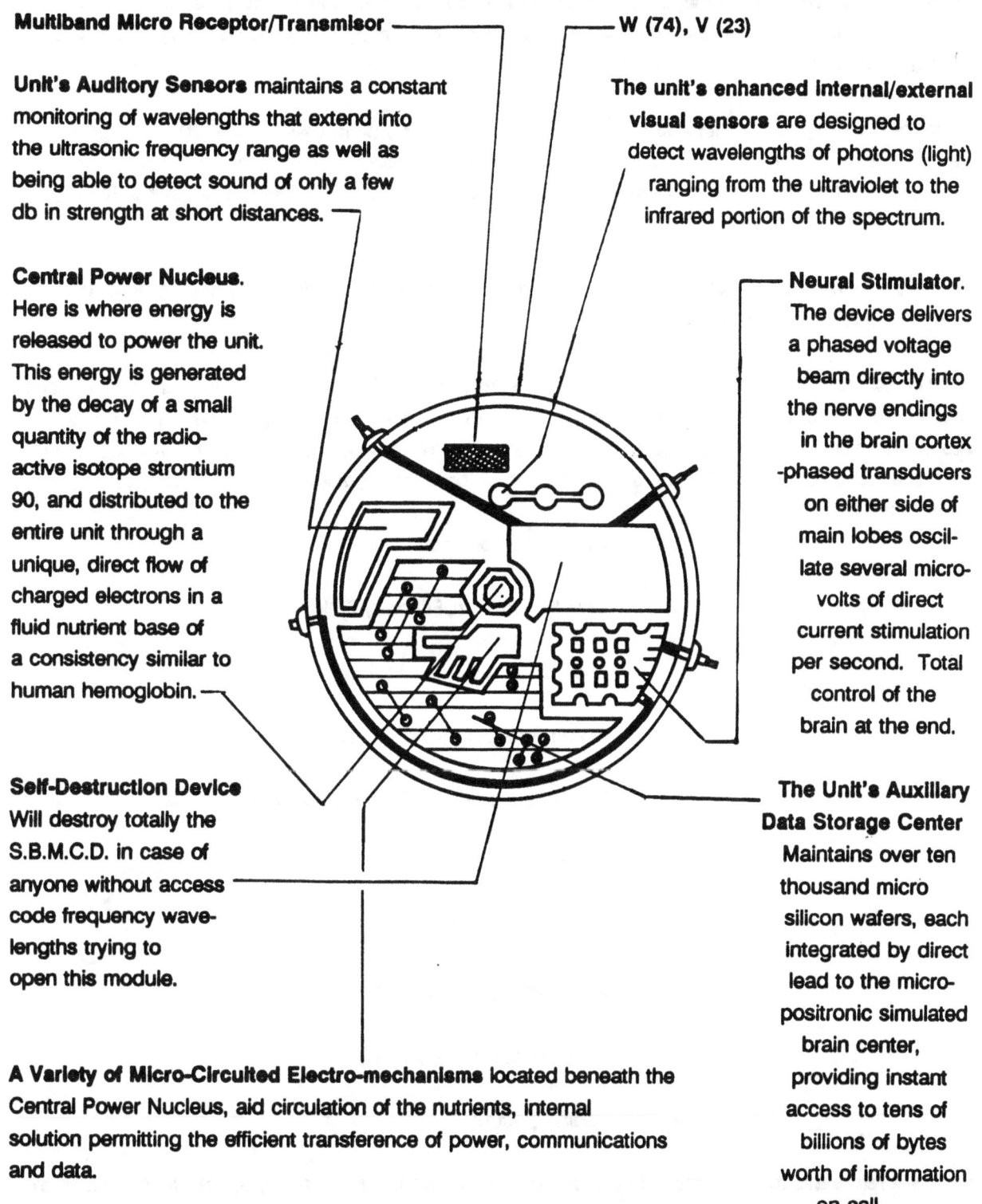

**Multiband Micro Receptor/Transmisor**

**Unit's Auditory Sensors** maintains a constant monitoring of wavelengths that extend into the ultrasonic frequency range as well as being able to detect sound of only a few db in strength at short distances.

**Central Power Nucleus.** Here is where energy is released to power the unit. This energy is generated by the decay of a small quantity of the radio-active isotope strontium 90, and distributed to the entire unit through a unique, direct flow of charged electrons in a fluid nutrient base of a consistency similar to human hemoglobin.

**Self-Destruction Device** Will destroy totally the S.B.M.C.D. in case of anyone without access code frequency wavelengths trying to open this module.

**A Variety of Micro-Circuited Electro-mechanisms** located beneath the Central Power Nucleus, aid circulation of the nutrients, internal solution permitting the efficient transference of power, communications and data.

**W (74), V (23)**

**The unit's enhanced internal/external visual sensors** are designed to detect wavelengths of photons (light) ranging from the ultraviolet to the infrared portion of the spectrum.

**Neural Stimulator.** The device delivers a phased voltage beam directly into the nerve endings in the brain cortex -phased transducers on either side of main lobes oscillate several microvolts of direct current stimulation per second. Total control of the brain at the end.

**The Unit's Auxiliary Data Storage Center** Maintains over ten thousand micro silicon wafers, each integrated by direct lead to the micro-positronic simulated brain center, providing instant access to tens of billions of bytes worth of information on call.

**The S.B.M.C.D. or Implant** - The Spherical Biological Monitoring and Control Device (SBMCD) is a techno-organic enhanced synapitid processor powered by a micro-positron flow that controls or mimics the functions of the human nervous system with micro-relay's duplicating the operations of brain engram patterns.

Normally the Greys during an abduction make an insertion of a 3mm device (SBMCD) through the nasal cavity into more close proximity to the brain of the abductee and after implementing subliminal post hypnotic suggestions that could compel the abductee to perform some specific act within the next two (2) to five (5) years.

Normally the SBMCD only can be removed when the abductee dies, trying to remove it while the abductee is alive means certain death.

**Aliens and the Electromagnetic Spectrum** - After reading this document, there should be no question at all in your mind, that there are intelligences that can manipulate or materialize any kind of object into our dimension. Let's take a look for a second, at the electro-magnetic spectrum. As you know, our visual spectrum only makes up a small portion of the whole. Look at what spectrums are involved with UFOs:

- UltraViolet Spectrum
- Blue Spectrum (UFO entry Field)
- Cyan Spectrum
- Green Spectrum (UFO is Visible)
- Yellow Spectrum (UFO is in a Vulnerable Status)
- Red Spectrum
- Magenta Spectrum
- Infra-Red Spectrum (UFO is Departing)
- Heat Spectrum (UFO Field)
- Radio Spectrum

If you will relate this to cases that you are familiar with, as far as appearance, Spectrum Shift when in flight, etc. you will see the applicability of the above diagram.

**Alien Artifacts Used at Mutilations**

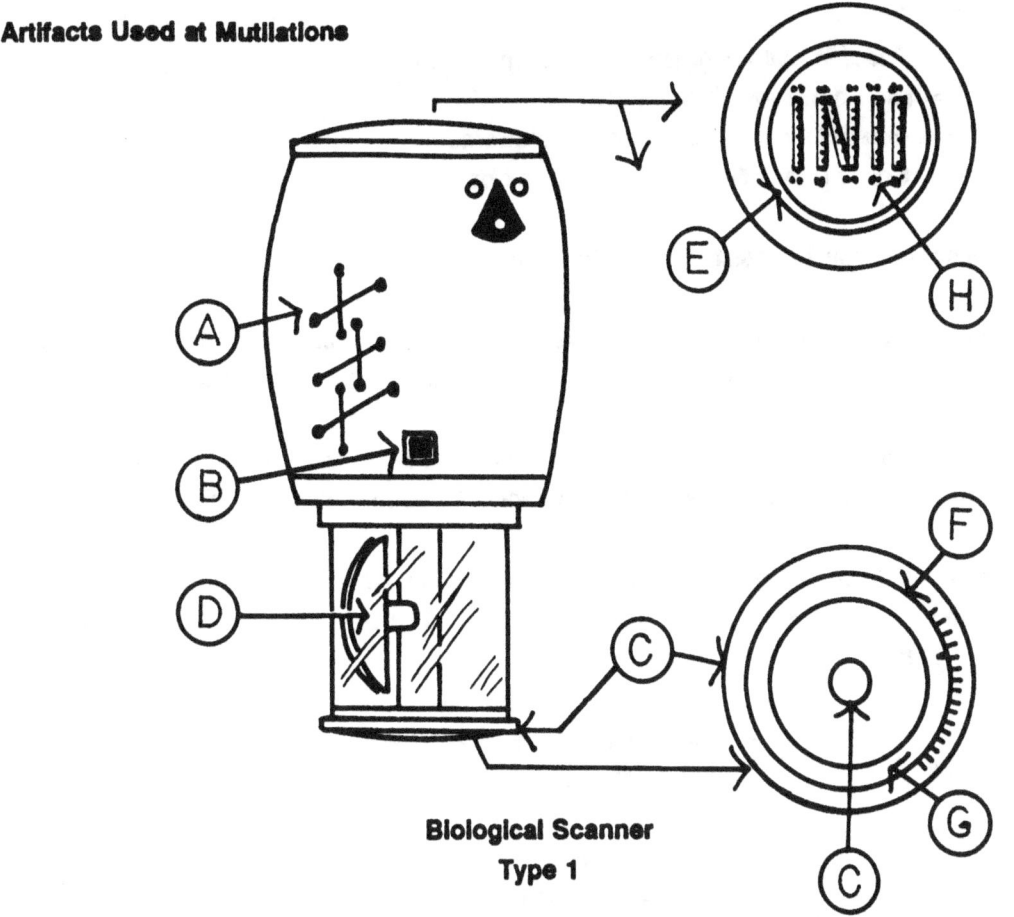

Biological Scanner
Type 1

This biological scanner is used for field determination of general medical condition of life forms by sensing body emanations in emission groups of Alpha, Beta, Gamma, Kirlian, Tetha and XI Rays.

OBS - refer to files of Majic medicine, advanced diagnoses practices, report for scale calibration and current interpretations of readings.

| Specification: | Overall Length | | 6.05cm |
|---|---|---|---|
| | Maximum Diameter | | 3,5 cm |
| | Weight | | 230 gr |
| | Sensitivity ($Em^v/AU^2$) | to | 0.0001 |
| | Reading Groups | | All |
| | (Alpha, Beta, Gamma | | |
| | Kirlian, Tetha, and | | |
| | XI Rays) | | |

A = Computer (circuits) housing
B = Activating Control Switch
C = Scanner Housing (#48 Mesh Screen)

D = Rotating Scanner
E = Reading Dial (increase from bottom to top)
F = Life Form Type, Select Scale
G = Ground Plane Screen (#48 Mesh Screen)
H = Select Range Scale for Different Kinds of Life Forms

The Biological scanner Type 1 consists of two (2) parts:

- Part number one is the scanner head, which you use to active the divice by rotating the scanner head, which is enclosed in a microchip mesh sensor network. This will activate the unit.

- Part number two is the main body, that interprets signals received and converts them into an audio readout, which is programmable by twisting the scanner head to the desired position.

The scanner functions much as does a mini-diagnostic center, by sensing the physical emanations of an individual or an organic being, which is in close proximity. The audio signals indicate either through its period or tone one of ten vital readings:

- Body Temperature
- Blood Pressure
- Pulse Rate
- Respiration Rate
- Basal Meta-Bolic Rate
- Cell Rate
- Lung Capacity
- Heart Activity
- Brain Activity
- Homeostatic Deviation

The scanner is intended for use with a single species, although it has limited diagnostic use on physiologically similar life forms. For visual readout, the scanner can transmit signals to a bio-computer, the lights on the computer serve the same functions as the oscillating tones, and can be enhanced for a more detailed appraisal of the being if desired.

NOTE: A regular physician must undergo extensive training before being able to "Read" precise information out of the sounds emitted by the scanner, and there is no simple method allowing the layman to do the same.

**The Spray Applicator** - The spray applicator is a small cylindrical dispenser of epidermal drugs employed by depressing the major axis of the unit between thumb and forefinger. Although it is capable of dispensing a variety of liquids, foams and gels, it is normally pressurized for field use with an organically based flesh-colored compound that acts as a coagulant to stop the flow of blood, as well as, an antibiotic to protect against infections. A mild local anesthetic deadens pain in the affected area. The applicator contains about five (5) to ten (10) doses, depending on the amount needed in field operations.

**The Field Reader Tube** - The reader tube is used when the low powered scan is not effective to diagnose a patient's condition, like a species that may possess a thick epidermis. It transmits life readings from a sensor head to four (4) independently, activated lights, which indicate through their intensity or period the following:

| | |
|---|---|
| Green | Heart Rate |
| Red | Pulse Rate |
| Blue | Body Temperature |
| Yellow | Blood Pressure |

The pointed sensor end of the tube must be placed into direct contact with the patient's exposed skin. Due to the narrow reading range provided by the reader tube, it is suitable for any Life Form.

**Surgical Scalpels** - Six (6) scalpels are contained within the Surgical Kit, Class: ABR-5, with cutting widths varying from one (1) to five (5) angstrom units. Activated by gripping the cylindrical base, they employ converging Laser beams for a very precise subcutaneous incision.

**Skin-Grafting Lasers**
The type one (1) low power Laser is used to quickly and painlessly heal external wounds by closing severed blood vessels and nerve endings, while stimulating the victim's anabolism, (constructive metabolism, i.e. regeneration of tissue).

The type two (2) laser can also be used to graft skin, (removed by scalpel from less sensitive parts of the body) onto areas where tissue has been damaged or completely destroyed. This procedure should be followed, as it is really necessary. The Laser is activated by depressing the dorsal bar increasing the pressure, causes and increase in intensity.

**The Blood Attracting Device** - This device is an emitter of a force-field, like some kind of Tractor-Beam emitter that can focus a wave or particle beam. The device is set for locate hemogoblin cells and attract them through the skin, using the principle of Gravitational Electromagnetic Focus Contention. This device can also be used for repelling, or mutual repelling and attracting mode to seize and hold stationary, repel, or attract even large objects.

The device can operate at a full 180/360 degrees, allowing it to function at almost any angle. Because the beam has its limits the wave particles are short-lived and will disintegrate into by-products at longer range, i.e. in excess of 100 meters.

# U. S. GOVERNMENT SECRET TECHNOLOGY

The helicopters themselves have been seen in areas where UFOs were reported, in many countries. In some of the more interesting accounts, the mystery helicopters were seen with UFOs or shortly after UFOs were sighted.

In a documented Mystery Helicopter wave in England, accounts place oriental appearing occupants in a unidentified chopper. Slant-eyed, olive skinned, oriental seeming occupants have been a staple at the heart and at the periphery of UFO accounts for years, significant numbers of the infamous "Men in Black, (MIBs)" have a similar appearance, but very often the are seen as very pale and gaunt men who are sensitive to light.

In "Stigmata" ne5, (fall-winter 1978) Tom Adams outlines the most prominent speculative explanations, accounting for the mutilation helicopter link, including the following possibilities:

- The helicopters are themselves UFOs, disguised to appear as terrestrial craft.

- The choppers originate from within the U.S. Government Military and are directly involved in conducting the actual mutilations.

- The helicopters are Government Military and are not involved in the mutilations but are investigating them.

- The helicopters are Government Military, and they know about the identity and the motives of the mutilators and by their presence, they are trying to divert attention to the possibility of involvement by the Military.

**Mystery Helicopters and Mutilations** - Normally the mystery helicopters belongs to the DELTA/NRO division. For a normal mutilation research mission they send one helicopter with a team of seven agents and two Remote Piloted Vehicle, (RPVs). The RPVs are small devices of look-a-like small saucers of three (3) feet in diameter and radio controlled.

**Surveillance Devices**

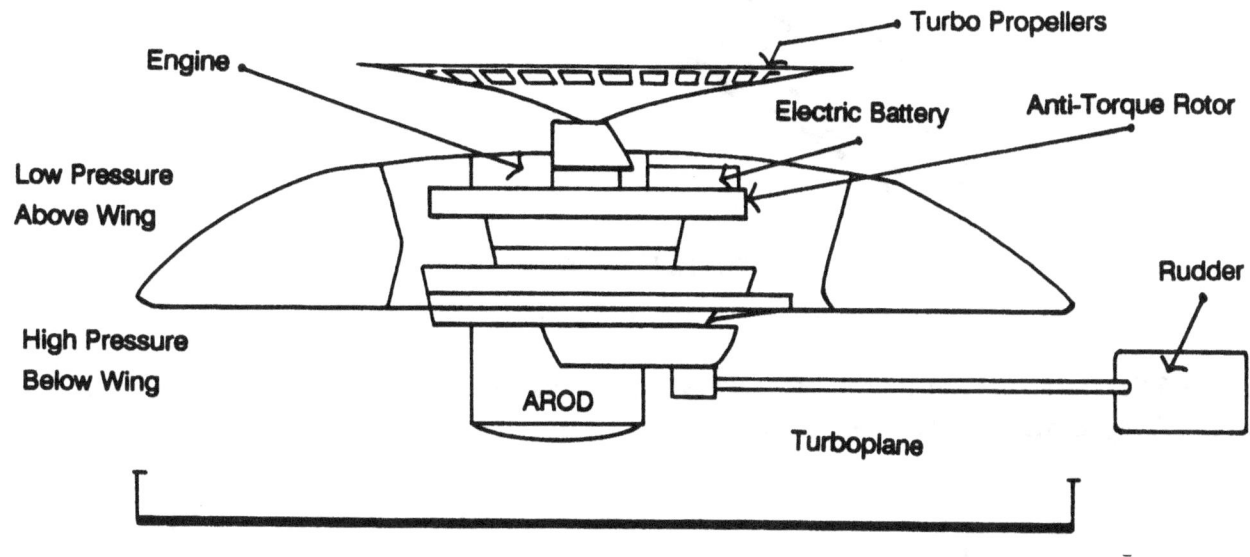

Airborne Remotely Operated Device, AROD
**THE RPV**

**NOTE:** This vehicle was created by Aerojet Electro-Systems in California.

The RPVs also carry a surveillance robot Airborne Remotely Operated Device, (AROD). A flying pizza with a tail, three (3) feet across, carrying a high-tech video camera aboard. Also surveiling the surveillance device from above, is a device known as a High-Altitude, Long-Endurance Drone, (HALE). HALE is designed to reach altitudes up to 100,000 feet. The device is powered by microwaves from the main helicopter.

**The HALE, ADR/238F**

The Teledyne Ryan Aeronautical corporation is responsible for the mysterious Helicopters or "XH SHARKS", as they refer to them.

**The TRA, SN-75
XH-75D**

I believe a cover-up of the capacity of this helicopter, is operating which overrides the real and total capacities and capabilities of this helicopter, but the following is only what I've been able to learn, and may not be the total capacity and/or capabilities.

- Type Rotor: Advanced Blade System, Experimental Research Demonstrator.

- Engines: One (1), 1,825 horsepower Pratt & Whitney engine, Canada J60-3A Turbojet, Turboshaft and two (2) 3,000 lb. (1361Kg) Pratt & Whitney Turbo's.

- Dimensions: Diameter of three (3) blade, double main rotors are thirty six (36) feet 0in (10-97m) length of the fuselage forty (40) feet.

- Performance:

  | | |
  |---|---|
  | Maximum Speed (level flight) | 276 MPH (445 Km/h) |
  | Rate of Climb | 5,000 Ft./Sec. |

Some helicopters (XH-75D Type) also have a Gravitational Electromagnetic Projector to create a "Cocoon" around the helicopter as a disguise. A 360 degree global arc with a radius of fifty (50) feet, more or less, provides a photonic force-field.

**A black Helicopter (XH-75D Type)**

When activated, turns into a UFO, as shown below, with no inertia and no acceleration effects.

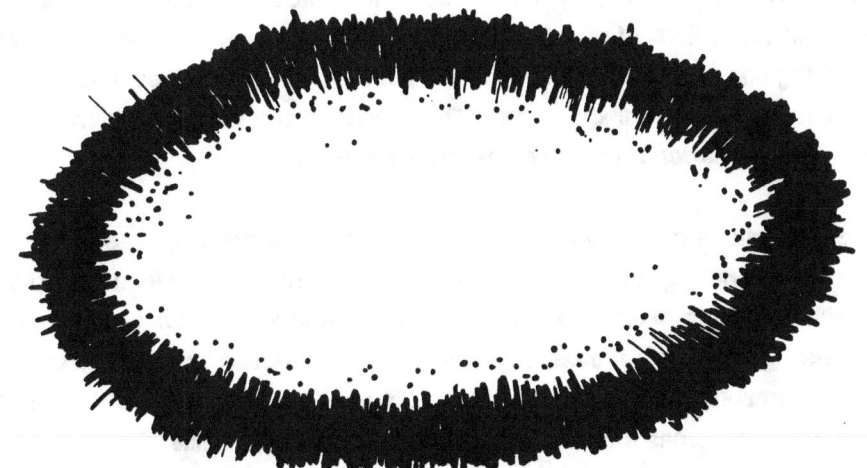

This provides 82% Higher protection against attacks. When used during a normal Mutilation Research Mission, the XH-75D Type black helicopter carries a crew of seven agents, two (2) RPVs, and one (1) HALE drone.

# ALIEN ABDUCTIONS

**"The Monitors": Abductions** - In the fifties, the EBEs, (Greys) began taking large numbers of humans for experiments. By the sixties, the rate was speeded up and they began getting careless. By the seventies, their true colors were very obvious, but the special group of the U.S. Government still kept covering up for them. By the eighties, the Government realized that it was too late and there was no defense against the Greys. So programs, (media, sine, TV, comics, commercials, books, magazines, cartoons, etc.), were enacted to prepare the public for open contact with non-human Alien beings. Now at the beginning of the 90's, (1991) these programs are continuing, and working well too.

The Greys and the reptoids are in league with each other, but their relationship is in a state of tension. The Greys only know the Nordics and the Reptilian Race as their Enemy, (don't confuse the difference between the Reptilian Race and the Reptoids or Reptiloids, because they are completely different races). We will talk more about the Reptilian Race later.

Some forces in the Government, want the public to be aware of what is happening, while other forces, (collaborators) want to continue making what ever deals necessary for an elite few to survive the coming conflicts.

The future could bring a fascist "New World Order" or a transformation of human consciousness, (awareness). The struggle is "NOW", when any active assistance is needed. PREPARE ! WE MUST PRESERVE HUMANITY ON EARTH !

**The Case For Alien Abductions** - GREY types and some NORDICs are on the top of an Alien Force who are bent on abducting millions of people (along our History) against their will, subjecting them to intense medical probing and committing other invasive acts. These determined Alien species are here, using Earth as a gigantic Breeding Ground. Their accumulated data show's that we're well past the phase where mythic or academic explanations alone will suffice.

Up from the strange, chaotic depths of abduction lore packed into a growing research files and journals, a central theme is emerging. The bottom line evidence of certain Aliens methodical, ultra-intimate program with Earthlings refutes the cosmic altruism found in traditional contactee literature, and counters the New Age trend of unrealistic idealism that concentrates exclusively on benevolent, spiritual motivations of presumed ET visitations. Scientists are beginning work on the Alien front, even the term "Extra-Terrestrial" stretches the data far beyond its reasonable limits. This unknown agency exhibiting strictly cold and clinical attitudes toward human beings is unquestionably "Alien"; it's certainly been witnessed often enough in conjunction with UFOs or UFO occupants. But even lengthy witness accounts that go into great detail about this or that Alien home planet or star cluster fail to factor out the elements of wishful thinking, deception and disinformation to the field of study. The potential for misleading data attaches itself to abduction research as readily as memories of blindingly bright lights and "missing time" recur as acknowledgements of the experiment itself.

The bottom line cannot be discerned beyond the facts at hand. Researchers admit to being stuck in the very earliest stages of divisiveness and confusion, more than anyone else. The abductees themselves have to feel the pain-terrifying literal pain that comes from being at the mercy of an Alien Force whose bottom line agenda largely ignores the fundamentals of human dignity and well being. But the Aliens do fully comprehend and seem to manipulate the requirements of organic reproductive creation. In this, they are much like us. We are dealing in a realm that overrides long standing cultural habits of thought and behavior. All we have to work with are the reports that have been allowed to break through the rigid veil that hides UFO information. The bottom line is babies !!! Time after time, abductees recall the same sequence of events and observations, their mechanisms of conscious control softened under hypnosis. Many come in "Clean", that is, with little or no previous interest in UFOs and having never heard of or much less read the several books that are introducing the topic to the masses: INTRUDER AND COMMUNION.

The proscribed order of abduction events reveals an industrial sized operation with across the board demographics. People of all races, most ages and both sexes have had the experience. It's fairly obvious, the researchers maintain, the manufacture of hybrid babies dominates the Alien Agenda. It's very extremely structured; they may do a "Million" different things, but they're really interested in one thing alone.

The generic sequence of events as recorded and documented, is as follows:

- Examination
- Egg Harvesting or Sperm Sampling
- Baby Presentations (Only for Females, rare for Males)
- Machine Examinations
- Media Presentations of Idyllic Environments. The abduction sequence is not exact in every instance, but enough leading indicators must be present to qualify the case as one deserving further investigation.
- Capture
- Examination
- Conference
- Tour
- Other Worldly Journeys
- Theophany - The delivery of some sort of quasi-religious message or experience.
- Return
- Aftermath

This hasn't happened in that many cases, but still deserves particular mention at the moment. A significant number of abductee's tend to retain special, almost spiritual feelings about their experiences, even if they were treated roughly or hurt. Also, recent investigative efforts are hinting that various sources in the intelligence community exhibit an odd preoccupation with the religious angle. These sources claim they can furnish an "inside track" to greater levels of truth about the UFO Phenomenon, and some investigators consider them trustworthy. They're inexplicable, "Fascinated" with religious side of the things. For some "nasty Alien Races" religion means the better way of mass manipulation and psychologic/sociologic control that they can get, and it works very well!

Researchers feel they are getting closer to having more definitive answers about abductions. A body of evidence that once emphasized matching details, lie-detection and the relative psychology of witnesses is giving way to one yardstick that never fails to command scientific respect. Similarities in form and content are important to know, but ultimate answers demand exterior evidence to bridge the gap between report and reality. After a close encounter of the 2nd Kind. We have documented the physical and chemical evidence such as soil samples and performed full analysis on these samples. The affected soil was packed in hard, desiccated chunks. To duplicate the effects, analysts had to heat normal soil at 800 degrees Fahrenheit for six (6) hours. This is one kind of physical evidence. Body scars and other medical trauma left in the wake of Alien examinations are perhaps the most incontrovertible evidence we found available. But other developments add more fully to the case for Alien reality. One example is an examination of stained clothing retrieved from an abductee just after her experience. The clothing awaits spectrographic analysis. The results of this analysis can tell us what kind of residual stains can appear, this again is additional physical evidence.

If large numbers of doctors in conventional gynecological practice were to be exposed to abduction data, a whole new sub-specialty would be inevitable. Many women abductees indicate that the Aliens tampered with their reproductive organs, if not actually committing a form of technological RAPE. Some women abductees are plagued with stark memories of strange wispy haired offspring whose presence sometimes evokes revulsion and the paradoxical feeling, "That's MY CHILD".

Additionally, we have to talk about the disappearing pregnancies. Abductees as far removed from one another as Brazil to New Hampshire tell nearly identical stories of pregnancies that suddenly appear and disappear and, later, reveal memories of repeated meetings with small, not-quite human infants they are asked to hold or touch.

One theory holds that the Aliens are attempting to learn by proxy the skills of nurturing to aid in the replenishment of their race. I think it's purely physiological. In some way we don't understand the baby is aided and/or helped by the physical touch. One of the abductees said she felt like a "Human Battery Charger". Elaborate rituals that present babies to parents and attempt to institute some sort of ongoing parent-to-child relationships have been documented, along with descriptions of high-tech nurseries and what looks like unusual incubation apparatuses.

The realm of UFOs is a place where paradox is the norm. It's assumed the Aliens use technology that far outpaces what we have on Earth, yet some of their methods seem low-tech by comparison. According to some data collected about abductions, the Aliens, like humans, have to confront the "World of E.r.r.o.r." in their experiments. For many years, the "World of E.r.r.o.r." appeared primarily as a crashed saucer, recovered pieces, and occupants....., what we're finding in the abduction phenomenon is that the factor of error is quite different. What it boils down to now is someone "sneezing out" something, or an odd metallic object in one's body that no one can make heads nor tails or. Several people have in fact sneezed out objects which may be Alien implants, through so far most of the objects have been lost or thrown away, but a few implants were recovered (which will be discussed later in this document).

The unfolding drama of abduction wouldn't be complete without a token note of "Big Brother" Government paranoia. Even earthly high tech includes sophisticated mind control techniques and the capabilities to project sights and sounds at a distance. Perhaps not probable, but it's possible that what most researchers consider an Alien orchestrated abduction sequence is actually a carefully scripted, electronically induced perception. That means big news to Earth:

> If certain areas of the human brain can be remotely stimulated, then it is possible to develop a technology that broadcast over large territories, literally saturating the area with a flood of symbols.

Such a device would be a major tool or social change. The buzz that human witnesses often hear could be produced by microwave radiation, known to be in the arsenals of governments darker programs.

A mind control hypothesis could also explain the relative lack of physical evidence, through only an irrational stretch of the imagination might link mind control techniques with what evidence has been documented, i.e. physical traces, unusual metal garments and scarring. Worth noting, however, is the fact that some researchers pursuing this trail of investigation, have been summarily and sometimes harshly thrown off. Sporadic accounts of humans assisting with abductions add a chilling note. What we inevitably face, even with the growing evidence, is a mountain of unanswered questions. If and when the mainstream quits denying the presence of UFOs, a well supported program of tactical research may ensue. Until then, abduction remains a mystery and the abductees the outsiders whose physical and emotional well being hinges on its solution.

The data we have about abductions, as derived by conversations with the abductees, that during the abduction process all abductees were scanned with a very sensitive and powerful mental probe which retrieved all personal and private data (name, address, age, profession, psychological pattern, etc...) from the abductees and was readily available to the Aliens for their data bank. Just a few of these names and addresses go to our government per the Alien/Government agreement. During the process of abduction a directional subliminal mental command, kept away any other non-interesting witnesses around the area of the Alien craft.

After an abduction occurs, and if we can maintain syncronicity of finding the abductees within 12-24 hours after the abduction, (really not an easy job), we recommend (in the case of a female abduction) a complete Gynecological procedure common to abductions.

In succeeding generations of family exposure to recurrent abductions, we are observing areas of the human body that are consistently and quite visibly affected. This evidence is related to the skin (dermatology) and muscles (musculo-skeletal system). During their experiences, many abductees and/or witnesses will feel a tingling, prickling, or static electric shock type of paresthesia sensation over the skin, followed by paralysis (involving the musculo-skeletal system) of the entire body, with the exception of an abductees heart and lungs where minor or no effects may be found.

**Scar's** - The body marks may remain permanently or become transitory in nature over a short duration, healing or disappearing all together. In dealing with the skin we find the following evidence:

- A thin, straight, hairline cut, linear, and about 1 to 3 inches in length.

- A circular or scoop like depression, about one eighth to three quarters of an inch in diameter and maybe as much as a quarter of an inch deep.

**Rashes** - Rashes are seen on the body, most appearing on the upper thorax (chest) area and lower extremities (thighs and legs). Many are geometrical in shape, triangular or circular. Other rashes, similar to chronic inflammations such as localized psoriasis, may be found on other body areas. First and second degree burns have been sustained in a number of cases and in some cases questionable tumors (Lipomas) have been noticed just beneath the skin.

**The Most Common Areas Examined by the Aliens** - Other medically determined patterns of consistency are as follows:

- The nasal cavity
- Ears
- Eyes
- Genitalia
- The umbilical (naval) Region on Females only.

These areas appear to be the physical areas of greatest interest to the abducting Aliens. Many abductees have described a thin probe with a tiny ball on its end being inserted into each nostril, usually on the right side. The abductees are able to hear a crushing type sound as the bone in this area is apparently being penetrated. We believe this is when they insert a device for tracking and communicating with the abductee in the future is inserted. Many abductees will have nosebleeds following these examinations.

**AS A PRECAUTIONARY NOTE:** We recommend that known or suspected abductees, who are parents, watch their children for any evidence of recurrent nose bleeds that can't be explained. I recommend immediately taking the child to a pediatrician to discover the nature of the nosebleeds.

**Diagram of Female Areas Examined**

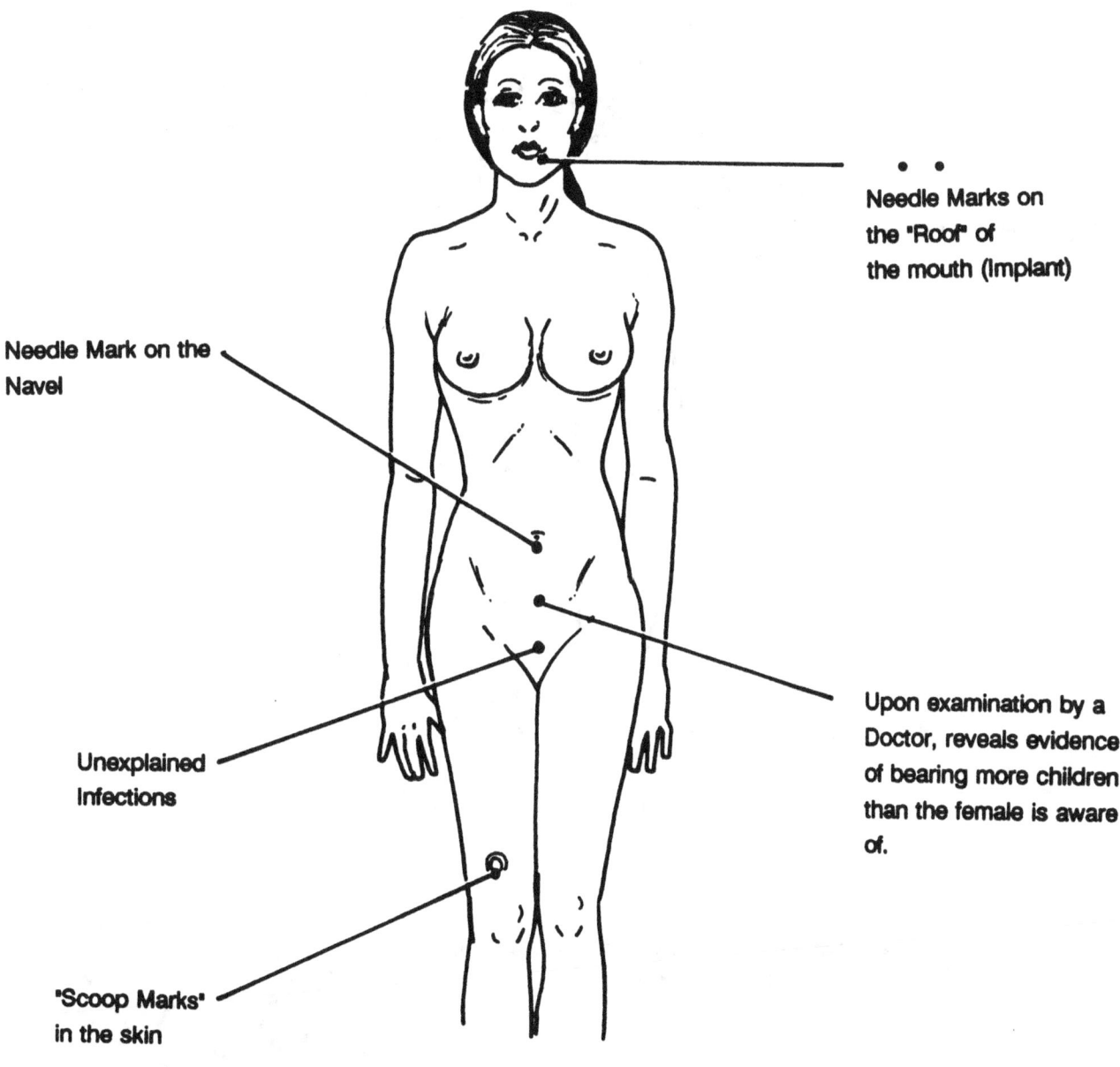

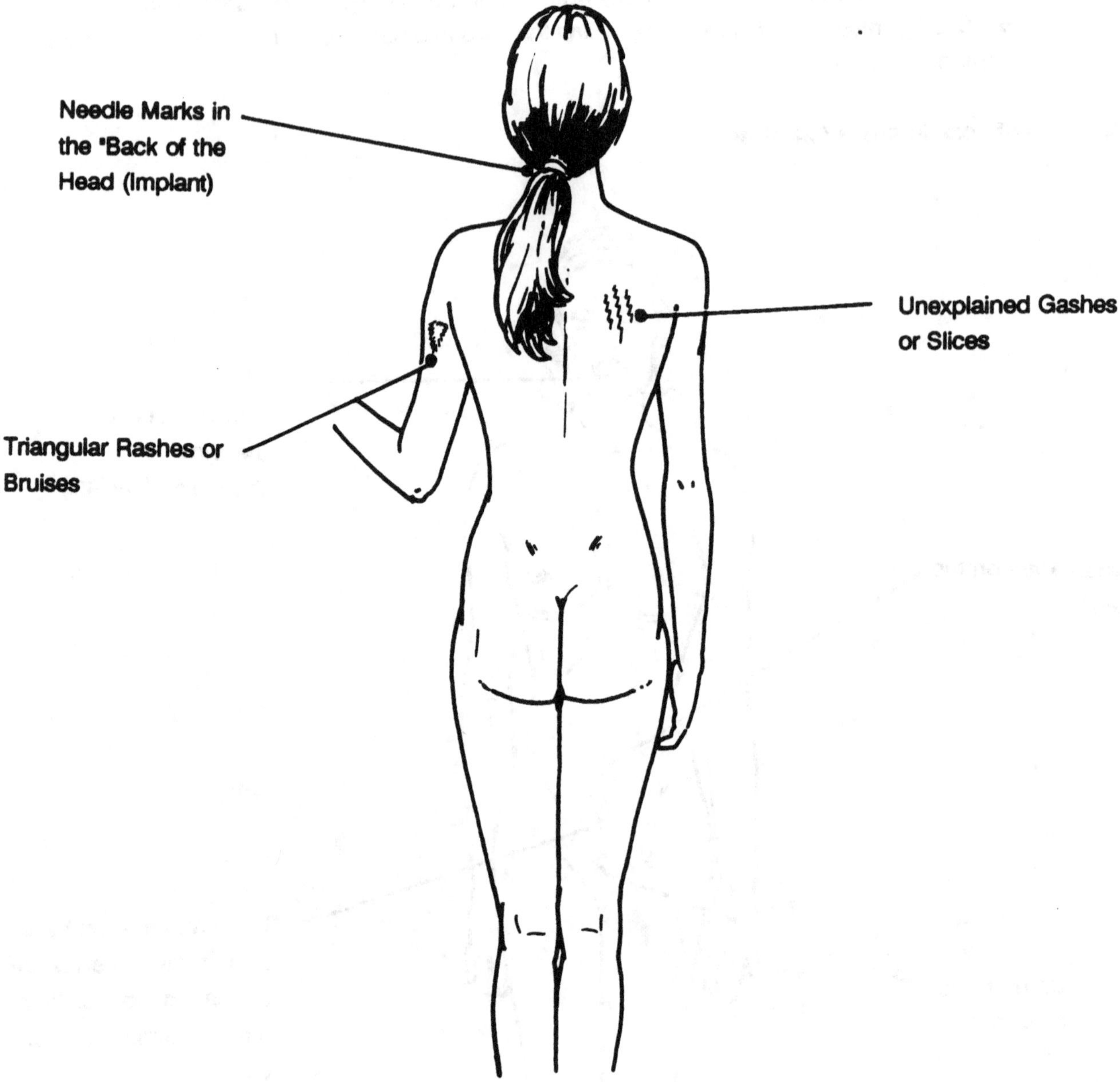

**Needle Marks in the "Back of the Head (Implant)**

**Triangular Rashes or Bruises**

**Unexplained Gashes or Slices**

Note: These are just a few of the locations, shapes, and types of Marks left by an abduction.

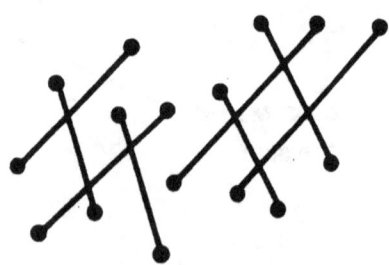

Many researchers believe that the Alien technology is being used to insert an implant (we talk about implants in other parts of this document) into this area for future tracking of the individual. It is interesting to note that many of the individuals subjected to nasal probing now have a future history of chronic sinusitis.

Documented evidence has also shown that some abductees have been probed in their eyes and ears with a similar instrument. With eyes being involved, abductees may experience temporary blindness, blurred vision, swollen, watery and painful eyes (photophthalmia), acute conjunctivitis (red and irritated, inflamed eyes that is called "Pink Eye" in lay terms). There's also some questionable history of these individuals developing cataracts.

Scars have been observed on the calf (including just over the tibia, or shin bone), thigh, hip, shoulder, knee, spinal column and on the right sides of the back and forehead.

**Biological Specimens, Samples Usually Taken from Abductees and/or Witnesses:** Evidence indicates that Aliens have taken blood, oocytes (ova) from females and spermatozoa from males, and tissue scrapings from their subjects ears, eyes, noses, calf's, thighs and hips, when abductees and/or witnesses are asleep or possibly under some form of Alien anesthetic.

There is also some circumstantial evidence to suggest that specimens might have been taken from the following; saliva, aqueousvitreous humor (eye fluids), cerebrospinal fluid, urine, stool, hair and nails.

We feel that all abductees are given some type of preparation prior to their examinations. Some witnesses have reported receiving "Oral Liquid" medication, others an application of a liquid solution similar to a pre-operative "prep" over various parts of their bodies; some report a tranquilizing effect "telepathically" transmitted from the acting Alien examiner, and/or application of an instrument to the head which renders deep relaxation or unconsciousness.

Next we will outline that what we think are the three (3) stages of examinations which abductees are subjected to (through some type of pre-operative anesthesia):

**First Stage - Pre-operative** Abductees are subjected to some type of twilight sleep state, where they're in a definite trance or daze

This twilight sleep state could be induced by several different things, from the liquid application over the body, a specific conscious suggestions by the Aliens, the Aliens using some form of our hormones or enzymes to stimulate a neuro-chemical response, or some type of yet unknown technology.

One odd note, in connection with this stage, that prompts a lot of questions, is why is it that so few abductees remember removing their clothes? Because during the process of "light sleep state" they receive a subliminal command to erase the more traumatic moments of the abduction. Some commands are so strong that just a few of the abductees are able to remember clearly their experiences, even when placed under hypnosis to remember.

**Second Stage - Procedures** Physical examinations take place; such as probing, insertion, exploration of the body, taking of biopsies, blood or skin samples.

During this phase, the abductees may be semi-conscious as the procedures are carried out. Some actually experience pain. Despite objections, the Aliens appear to be indifferent to their victim's pain and suffering; on the other hand, some abductees at this stage are given heavier sedation to quiet their fears and apprehension and do not recall any pain with these procedures.

**Third Stage - Post Operative** Afterwards, the abductees and/or witnesses say their bodies feel sore or exhausted as if having been involved in strenuous activity; some explain that it feels they've been tossed around or "hit by a Mack Truck".

This is similar to the known effects of "Curare" a drug originating in South America that induces therapeutic muscular paralysis. There are specific blood enzyme studies, that if performed within 12-24 hours of exposure, can be used to detect any abnormalities of muscular activity.

Several documented physical traces of the abductee experience correlate fairly well with some of our recognized medical procedures. Most outstanding is one called "Laparoscopy", which is a cylindrical, tube like instrument with special optic attachments are placed through a females umbilical (naval) region for exploration of female organs. With this particular instrument, a physician is able to observe all female organs to determine if any abnormalities are present, as well as obtain OVA-EGGS from the ovaries. Most women abductees have felt they were being "blown up" inside, feeling tremendous pressures in the lower abdomen and discomfort in the vaginal area. During the Laparoscopy procedure, approximately two liters of carbon dioxide is instilled into the abdominal cavity. This causes distension of the abdomen, thus allowing better visibility of female organs. A few women may have residual scarring from the long, needle like instruments placed through the umbilical area.

**Pregnancy** - In correlation with the Laparoscopy procedure is a new treatment for infertility called Gamete Intra-Fallopian Transfer (GIFT), which treats infertiility by placing sperm and pocytes directly into the infertile womans fallopian tubes for in vivo fertilization. In contrast with In Vitro Fertiliization (IVF), GIFT facilitates natural physiologiical processes to achieve pregnancy.

Male abductees report having a tube like device attached to the penis which causes ejaculation for sperm sampling; this is highly uncomfortable to the individual. Most have said they sustain transitory small lesions that disappear shortly afterwards. Others claim to have experienced direct sexual intercourse with an Alien Hybrid Female, ostensibly for the purpose of sperm retrieval as well.

A substantial amount of evidence has been accumulated that female abductees undergo gynecological insemination. These events have resulted in pregnancies, most documented by a positive pregnancy test by gynecologists. In cases of spotting of bleeding with these pregnancies, the physician normally performs a pelvic Ultra Sound scan (Ultrasound test to check pregnancy within the uterus) to rule out a threatened miscarriage or missed abortion (failure of the pregnancy to grow). The test can detect a pregnancy as early as five (5) weeks. Thus, pregnancy can be confirmed with two different gynecological tests. It should be noted that the government uses the abductees normal doctors to collect most of above and following data, without the patient knowing anything, and in most cases the doctors knowing very little, and threatened that if they ever talk, they will have the licenses revoked "permanently".

In cases we have seen, this is followed up by a subsequent Alien abduction, usually when the pregnancy is approximately 8 to 10 weeks along. During this time, there is a retrieval and/or removal of the pregnancy by the Aliens. The female may or may not experience spotting or bleeding during this time. What's baffling is how such a pregnancy can be removed intact, without causing death or injury to the extracted embryo by the Aliens.

In order for a particular to continue, several methods would have to be available immediately to sustain life in the extracted embryo. We suspect that some type of simulated intrauterine growth incubator is used to maintain the pregnancy; or perhaps an Alien Hybrid Female is used, whose role is to act as a surrogate type mother. Techniques for re-implanting the embryo are very difficult for our medical technologies of today.

Several alternative combinations can be speculated with the artificial insemination of female abductees:

    A. Alien sperm plus female eggs
    B. Male sperm plus female eggs, (in this case the Aliens choose a special type of Male).
    C. Male Hybrid sperm, (combination of both A and B) plus female eggs.

NOTE: Female human eggs are apparently needed in all cases.

- **Hybridization** - They claim that RH O- blood is the proof of hybridization and our own science tends to bear out their claim.

- We know that this group of Aliens has a tendency to lie. Since this is a "device" (reffered to above) is of their invention, they are probably able to manipulate it just like Hollywood technicians manipulate equipment to produce the "Special Effects" or trick hologram photography.

**Our Genetic Code Is Under Siege** - In focusing on the gynecological and reproductive procedures that have been performed on abductees, we have to come to firmly believe there is some type of ongoing genetic manipulation that is occurring within various family generations. For purposes of clarification, it is essential for us to use some medical terminology to explain specific facts. The way to genetic manipulation lies within the Deoxyribon Nucleic Acid (DNA) molecules of the human gene cells. These genes control the reproduction and day to day functions of all cells. It has been estimated that there are probably 30,000 to 400,000 essential genes in the human cells, assembled in lengthy linear arrays that together with certain proteins form rod-shaped structures known as chromosomes. Chromosomes from certain individuals, through certain altering techniques, form a customary arrangement or "Standardized Format" known as Karyotype.

We believe it is the manipulations of the genetic codes on the Karyotype of individual chromosomes that the Aliens are researching, specifically looking for Mutation Patterns. Mutation patterns would allow them to re-arrange the genetic coding on certain LOCI Lin Genetic terminology, the specific site of a gene in a chromosome.

In this manner, they would be able to experiment with a multitude of LOCI in the various chromosomes, thus bringing about new genotype individuals in proceeding generations. Perhaps each succeeding generation of families is subjected to a different comparable type of experimentation by the Aliens.

It is interesting to note, just where the residual scars are located on abductee bodies. Now I would like to pose a new interpretation of the evidence found in our ongoing research of abductees. Many scars are found over the shin bone (tibia) and the hip bone (iliae crest), which are common areas for obtaining bone marrow samples and/or aspirations. Simply speaking, bone marrow produces the red blood cells in our bodies, and it is significant, in our opinion, that it is through a humans bone marrow and blood that it is possible to study their individual chromosomal pattern.

The timing of abductees first experiences also correlates to the hypothesis. The marrow of essentially all the bones in the human skeletal structure produces red blood cells until a person is 5 to 6 years old. The marrow of the long bones, except for the proximal portions of the humor, (upper arm bones) and tibiae, becomes quite fatty and produces no more red blood cells after the age of twenty. Beyond that age, most red blood cells are produced in the marrow of such areas as the vertebrae (Spinal Column) and sternum (Chest Bone), ribs and hip bone (Ilia), and as we get older, these bones become less productive.

Most initial abduction experiences occur when the abductee is between the ages of five (5) and twenty (20) years old. Based on these observations, we are probably dealing with two very important phases of an abductees life as it relates to this apparent Alien exploitation:

### Pre-Adolescent/Teen-age Phase

Initial blood samples, bone marrow aspirations and tissue samples are taken between ages five (5) and twenty (20) years of age. This would be the time frame for specific and early genetic studies on the chromosomes of abductees as a follow-up from another generation with studies performed to see if a certain pattern is consistent within the particular family. During this time, some type of implants are inserted so that the individual can be followed and found at some future date, after further analysis of samples have been obtained by the Alien researcher. Some of these individuals may be abducted again, possibly due to a failed implant or for re-confirmation on certain genetic information.

### Adult Phase

The pattern we are seeing at the present seems to reveal abductees are undergoing some form of ongoing genetic exploitation, implementation or manipulation, such as genetic coding. This is where we see the most consistent, most documented procedures of the abduction phenomenon, such as artificial insemination techniques, which are continued on some abductees throughout their reproductive years.

- Perhaps in their own way, if you believe in the "Space Brother" theory the Alien scientists are meticulously, methodically and clinically re-structuring the human race through procedures of genetic manipulation, so that we will be "designed better" for resolving our own problems, rather than requiring their more direct intervention into our petty world affairs.

- Maybe there is a double-blind study on the human race ongoing, for reasons known only to the Aliens, who are running the program.

- Or consider the possibility of this being an Alien scientific research mission, where studies are being performed on technologically less advanced planets.

All of this is pure speculation and wide open to other theories or hypotheses.

A new eugenics movement has recently sprung up in Southern California. Eugenics is the science dealing with improvement, by control of human mating, of the hereditary qualities of a race or breed.

These issues that bear on the eugenics movement can be argued both pro and con. However, we feel that overall it is a very emotional issue with implications that could correlate with the Alien agenda, and either certain problems that could very well rival our nation's controversial abortion decisions.

The techniques of selective breeding which seem to be utilized by the Aliens are beginning to appear in our own culture. For example, Dr. Robert Graham, who founded an organization called Repository for Germinal Choice, has come under heavy attack by his adversaries. Essentially, Dr. Graham's program allows women to "shop" in his repository (Sperm Bank) for a specific type of donor sperm that they can retain for later use in artificial insemination procedures. In other words, a woman could select the sperm of a Nobel Prize Winner, an Astrophsicist, or an extra-ordinary professional such as an Astronaut, Physician, or a Mathematician, etc. It appears that this organization is out to create a new generation of super-intelligent children who will be eventually adults, through the techniques of selective artificial insemination. The socio-economic, political, psychological and theological implications of this controversial new technique, whether put to use by Aliens or humans, go beyond the scope of this document. Exactly what are these implications, you might ask? Only time will tell!

**The Metagene Factor** - The Metagene is a biological variant lying dormant in select members of the human race, until an instant of extraordinary physical and emotional over-stress activates it. That's an energochemical, in response to adverse stimuli. A chromosomal combustion takes place, as the metagene takes the source of biostress, be it chemical, radioactivity, or what ever and turns the potential energoresponse into a catalyst for genetic change. The main focus of the catalyst power is a gland in the middle of the human brain called the PINEAL gland, and the nutrient for increasing the Pineal's action is the adrenalin. The metagene factor gives the ability of Psionic Power.

The main interest of the Aliens especially the Greys is to understand and control the Metagene for their own race. They try to do this using Biological Experiments to make Hybrids from both humans and Aliens.

They believe perhaps the men from the planet Earth are the deadliest creatures in the Universe. Because only on Earth people are apparently capable of generating the Metagene Factor, which means Natural Psionics ability, "Real Power". The principle races in the Universe are psychologically the same. The pure cold logic is a normal order to most important races. Basic sameness makes for predictability and security, the enemy one knows, are the ones you can guard against. This is not the case with mankind.

While most are uniformly human, some, many more, apparently, than anyone had dreamed possess a latent tendency towards super humanity, Natural Psionic Abilities. That in itself could prove dangerous for any idea of Alien domination on Earth. But couple of mankind's inherent belligerence with the fact of the Metagene affects, each human is unique and Earth becomes a spawning ground for a unpredictable super-race, "if we have the chance".

Others have already demonstrated an awareness of Man's Potential along the human history, (Nordics treat). This is because the Aliens are here to try to control Earthlings before we dominate them, and they want our most important secret: THE METAGENE FACTOR, which is the Aliens only hope.

## ABDUCTION BY GREYS

There is a large number of Grey abductees living all over the world. Most are either unaware or just partially aware of their involvement with the Aliens.

Grey abductees seem to be taken for a large variety of reasons, some apparently having nothing to do with the person, him or herself.

They might have been manipulated, to various degrees, for no other reason other than that they were with a person scheduled to be abducted. Others have been abducted apparently only once or very rarely, compared to throughout life. As with Nordic contactees, it would seem that genetics and artificial racial background plays a part in their selection processes.

People have been abducted by Greys for a long time. This contact can exist during generations in the same family, and even cases of the abductee having contacts with same entity during the process of past lives.

All this means, that there may be a clear indication that a large percentage of the human race could very well be a "Sleeping Army", and were picked up only once or twice and implanted, but not generally bothered or contacted throughout their lives. This suggests that these people are on hold for something, or that they are simply walking transmitters of some kind of information, basically bio-energokosmic psychowaves, back to the Aliens.

Most Grey abductees are first taken as children, and many adult abductees know or feel their children are being abducted. It would seem that most children who appear to be involved don't want to consider the possibility of actual abduction or think about it at all. It would also seem that the best policy with most children is to leave it at that, not push them to remember. Hypnotic regression, especially, is considered too much to ask of a child, even adults, actually should not take lightly the decision to have themselves hypnotically regressed to recall abduction experiences.

Children and adults alike, if they are involved, will tell you of their weird dreams and experiences even when they don't believe there is anything going on.

Women are frequently abducted largely for reproduction purposes, through some never experience that kind of interaction. Some of those who are involved in that part of it report that their earthbound, human children that "live with the Aliens" have several different reactions to the situation. Some are perfectly happy with the program as it is, many don't express any opinion about it at all because they don't see it as being anything they can do something about. Some are repulsed by the children and by their own involvement with them. Some have felt that the children we produce will be used against us.

Some of the weapons and aircraft training that abductees are involved in is actually done by the abductee for the sake of the unborn child. Training the mother as received is thought to be transferred psychically to the unborn child. Things along this line are part of what might inspire some abductee's to fear their halfbreed children. Others love their halfbreed children as they would any others. Many who feel this way would like to recover their children and bring them back to Earth where they would have a better life than with the Aliens. Many also feel that the children are better off where they are but would certainly like more open contact with them, more interaction with their children.

**Aspects of Bio-Field Differential**
The bio-physical field of the Grey entities is in opposition to the field of the human body. The reaction produced by the interface of the two fields can produce a sensation referred to as "body Terror" by abductee Whitley Strieber. A human may or may not experience this reaction, depending on their ability to develop a "Mental Block" to the reaction created by the field differential, on occasions, the same possible reaction with some Nordics is experienced. The fields emitted by the Greys parallel the type of field emitted by much of their technology. The Greys use technology to amplify the effect of their bio-fields on the humans. It allows them greater control over the ordinary humans, who may react to the field differential with the human emotion of fear.

**Using Internally Generated Tones to Interfere with some kinds of Manipulations**
Anyone can produce high pitched tones in his or her own head. It would render the manipulative technology inert. Just maintain the tone during the presence of the Greys and anyone is capable of arguing with them. Generation of the tone is an important part of the human defense programming against intruders.

### Awareness Parameters for Tone Maintenance
Since the technology used to manipulate humans encourages relative paralysis of conscious functions, maintenance of tone-interference or other mental techniques must be handled by a form of bridging, or cooperation between conscious and sub-conscious awareness. Effective use of counter acting techniques depends on an increasing awareness of both states. The sub-conscious areas are where all memories are preserved, and all experience there relative to abduction is subject to blocks placed there by both the abductee, (as a programmed response to a function which the Aliens have connected with the survival impulse) or the Aliens. Lucid dreaming indicates a closing of the gap between awareness of both mental areas. Being able to interact with the Greys, on their own turf, so to speak, is a threat to Alien manipulation and control.

**Effects of Multiple Level Alien Manipulation**
The most interesting aspect of Alien control and manipulation is that the technology they use allows the separate and distinct manipulation of both the gross and the finer densities that compromise the physicality of the human body. They have the ability to place a human on a table, put him or her into Delta Sleep, shock him or her with a static charge, separate his or her finer bodies, and manipulate

them at will. Manipulation of finer areas that involve the formative forces of the physical body have tremendous effects at the physical level.

They also have the ability to withdraw experience and memory from a human and place that experience and memory into another body, or container, whether the container is a natural or synthetic one.

Manipulations of these type have effects on the ability to have one's consciousness leave the body, emotional response patterns and programming, ability to deal with psychic blocks, and a host of other parameters.

To begin to counteract this, work must be done by the individual to work on expansion of awareness. Connections must be developed between the conscious self and the "second self", which may be the self that is projected astrally. After a while, the conscious maintenance of the connection begins to combine both conscious memories and memories which lie deeper. Work on self hypnosis and regression techniques with a qualified individual. These techniques will aid in the process of developing more awareness and control. Constant experimentation while in Alien presence must be done by the abductee, who must make an effort to note results of various mental manipulations on the Alien control efforts.

**Some Major Indicators of Alien Interaction and Possible Hidden Involvement**

- Missing Time
- Waking up during the night or in the morning with unusual bodily sensations, such as tingling, numbness, dizziness, heaviness, or paralysis. Any of these are often accompanied by disorientation.
- Nightmares or vivid dreams of Aliens and/or their technology.
- Sleep disorders, also waking up at a specific time.
- Physical marks or evidence or bodily manipulation.
- Repeated sightings of Alien craft.
- Clear remembrance of Alien contact and interaction.
- Healing or inexplicable improvement in your physical condition.
- Reactions of fear, anxiety or unusual bodily sensations upon viewing visual images of Aliens or their technology.
- Feelings of having had a communication.
- Un-explained behavior totally inconsistent with previous patterns.

Any or all of the above indicators may appear in combination with other ones. Many of the indicators could also signal a physical problem requiring medical attention.

Clinical psychologists have reported that a significant number of people have been helped by their interaction with alien species, mostly through Alien physical manipulation that resolves a physical condition.

**Bio-Conditioning of the Hybrid-Fetus in a Human Host**

The nature of the Alien physiology is quite different from that of the human host. Through the autopsies that we have done on the Greys, it has been found that the tissue is very often a composite that reflects a blending of animal and plant genetics. Some species have black tissues with green blood and appear to function on a light based nutrient system.

Many human hosts have reported mysterious infections, often yeast-like in nature. Through analysis of many cases, it has been pretty well established that the Aliens often purposefully introduce systemic organisms into the human female in order to acclimate the hybrid fetus to the physiological environment of the Alien life to come. These substances are thought to be micro-fungi and/or micro-viruses. The implantation of these substances is thought to be critical for some species in their hybrid breeding program.

It is quite apparent that here is no concern for long-term affects of hybrid breeding on the human. They apparently make adjustments during the breeding cycle. These adjustments can include substances which physically wear down the human host, as well as, manipulation of the various fields inherent with the human body.

The question always seems to arise, if these Aliens can produce synthetic tissue, why couldn't they just grow a torso or create a synthetic clone to bear the fetus for them. Why do they have to use humans for this process ?

There are two main parts to the answer.

1. The first one is that they need the genetic input. The whole reason for hybridization is to enrich the genetic capability of the organism, for this they need human genes.

2. The second part of the answer is that they need the effect on the fetus or human emotional and psychic experiences. The hybrid being will develop growing sentience, (unlike a synthetic, where the sentience does not extend past the physiologically determined by the matrix, the programming is set) and must be exposed to these factors.

One would then ask why they could not sit around and project psychic energy at a clone, (using artificial ways, of course). The answer is that the emotional component would be missing and the psychic energy not so pure.

The hybrid fetus is extracted from the human most about ninety (90) earth days and development then continues in a totally Alien environment.

The whole process causes premature deteriorization of the human female host, which simply increases the vulnerability to Alien manipulation due to a weakened condition. During the hybrid breeding process, the body of the human female host will be manipulated.

There have been cases where fillings in teeth, which may contain mercury and other metals, have been taken out of the humans mouth in order to seemingly promote the health of the developing fetus.

**Abductions Related Medical Anomalies, an Abductee may Experience**
- Abnormal blood cells
- Implants in the eye, skin, nose and/or anus
- Diarrhea, mostly in men
- Constipation, mostly in women
- Shoulder pain that comes and goes
- Lower back pain, three (3) vertebrae up, from the lower lumbar
- Knee pain, in the dip of the knee under the knee cap
- Constipation, mostly in women
- Rashes, immediately after contact, may be caused by radiation
- A Lump in the dip of the collar bone at connection to neck, may cause paralysis or semi-paralysis of arm or arms.
- Artery in the wrong place, may indicate a arterial implant
- Pregnancy connected with an abduction time period, usually a small fetus
- A baby may have an unusual appearance
- A child who has ESP and/or is advanced beyond his or her years
- An eyelid, which may contract and roll outwards under stress
- Eyes may be able to see through the closed eyelids
- Visual marks on a body
- Scars or geometric forms visually present on the body
- Scars on the back of the legs for people born in 1943, seems to indicate skin scraping, i.e., cloning
- Hear a buzz, beep or modulated tones in their ear, when going to sleep
- When they feel a "Last Breath Feeling", may be just a side-effect of a out-of-body experience
- Shaking of the bed, in conjunction with the limbs of the their body floating or rising
- Geometric symbols seen in the mind
- Tiredness in the morning after a good nights sleep. Feel as though you are worn out, sleep learning
- Out of body experiences
- Sudden development of ESP
- Wryneck, enzyme to the brain, (new development)
- Extra vertebrae in the neck
- Egg sized lump on the bottom rib
- Warts form a geometrical shape after examination, appear to be from some instrument.

# INCULCATION DEVICES AND
# THE PRACTICE OF MANIPULATION OF THE MIND
(The following material is copyrighted by Nevada Aerial Research)

**Learning, teaching and assimilation techniques of electronic space societies**
Through the experience of those who have had and are continuing to have various contacts with non-human entities, we have learned some data regarding these advanced societies. Overall, life appears to be rather regimented, social codes are sufficiently advanced beyond those of the human being, that they become a part of existence rather than the subject of existence.

In electronic space societies of positive orientation, knowledge and data are generally assimilated under duress, in that training and education are accomplished by forceful means, such as implant technology.

Techniques of electronic space societies inculcation, (to teach or impress by force or frequent repetition) is the usual term used.

One of the prime patterns of activity is what is called direct observation. During this process face to face contact occurs between the Alien species and a volunteer humanoid for the purpose of exchanging cultural information. Direct observation also appears to include various types of medical analysis. Abductees in this situation are actually volunteers, but inculcation methods block their knowing or awareness of their own participation. Of course this is not the procedure of some Alien races, who take abductees against their will, i.e. the Greys.

**Direct observance operational procedures** are as follows:

- A human target is sensed.

- The humans ridge response system is examined and checked.

- The craft is allowed to appear to the human, releasing old implant blocks.

- The subject is monitored and picked up when appropriate.

Sometimes a targeted human being will be picked up in childhood and taken to an underground facility for crystal implantation. They will be monitored through the growth period. In young adulthood, the human is picked up, and the crystals are removed and the human is kind of "Placed in Reserve" for future triggering and employment. The Andreasson Affair is an excellent case which features these elements.

## INCULCATION DEVICES

Several kinds of devices have been noted by those who have had contact with various Alien species.

**Inculcation Bar** - A rectangular box of metal with multi-colored lights displaying a sequence of blue-green-red-green-blue and the numerical sequence of 1-2-3-4-5-4-3-2-1. The subject lies on his/her back, on a metal table directly beneath the bar. At the foot of the table, level with the surface in a metal console box with cables connecting to the table. The attendants are dressed in red jump suits and the room itself is bathed in red light, except when inculcation therapy is in progress.

**The Inculcation Monitor** - This monitor, when coupled with the data banks in the ships computers, can be used to implement the enhancement of a being by infusing abilities through implantation of data. During this process, it appears as if negative mental states are removed and mental barriers are set aside. The monitor is a modulated catecholine encoded laser beam, which pierces the eyeball of the subject in a earths nano-second (one billionth of a second). It resonates down compatible optic frequencies, which sets up harmonic vibrations which disseminate the encoding to the proper receptor sites.

**Catecholine Beta-Lipotropin 4753** - is a technical name for a substance often given to the abductees. It has the effect of removing certain body neuronal blocks and at the same time gives a boost to the awareness/intelligence level of the abductee. It is a mixture and distillation of slightly enkephalinic melanocyte-stimulating adrenocortitropic hormone in a neural metastasizing, neurotransmitting medium; i.e., a cerebral cortey "Roto Rooter".

**Rapid Inculcation Processes**

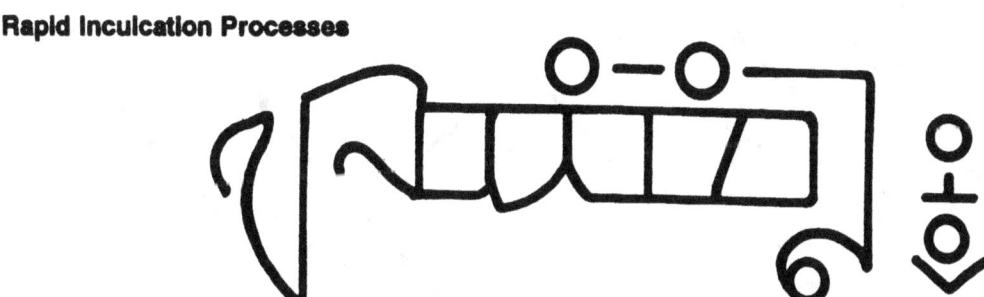

I one method, one dons a helmet bearing wires, (old Zeta Reticulae-2 technology, almost obsolete) and needles. A crystal cube is put into a niche on top and a type of strobe light flashes in resonance with the brain waves of the abductee. The abductees head becomes filled with pictures which gradually form a pattern of responses. The abductee is given a learned motion, pattern, response system, and may be trained in a brief amount of time to do a complicated task.

Sometimes the abductee is hypnotized or made to sleep and a high frequency microwave emission is used as a rapid inculcation method. Rapid Inculcation processes an occular wave with which to send encoded data into his or her nerve/ridge response system. This data may be triggered later by a pre-arranged stimulus response signal that will be present in the environment of the abductee.

Sometimes the bio-energetic field itself is used as a carrier wave and the data is encoded in sound-code-symbols. The Grey hybrids, (Rigelians and Beteloiustans) have an interesting variation on which the abductee sits in front of a screen and computer console type and interacts with images on the holographic display.

**NOTES:**

**1. The Ridge Response System** = A series of energy focal points formed by the energy streams around the physical body. Ridges can impinge on each other and cause an enduring state. Ridges exist in suspension around the human and are the foundation upon which facsimiles are built.

**2. Facsimile** = A facsimile is a recording of all perceptions, effort, thought, and emotion experienced by the organism, during any one point in time. It is recorded by the reactive mind. A facsimile can also be considered the physical universes impression on thought, specifically that section of thought which has a physical universe impression with a time tag on it.

**3. Reactive Mind** = A reactive mind, are portions of the mind which works on a stimulus-response, which is not under conscious control and which exerts force and the power of command over the beings awareness, purposes, thoughts, body and altiosis.

The reactive mind never stops operating. What is thought to be unconsciousness is always conscious.

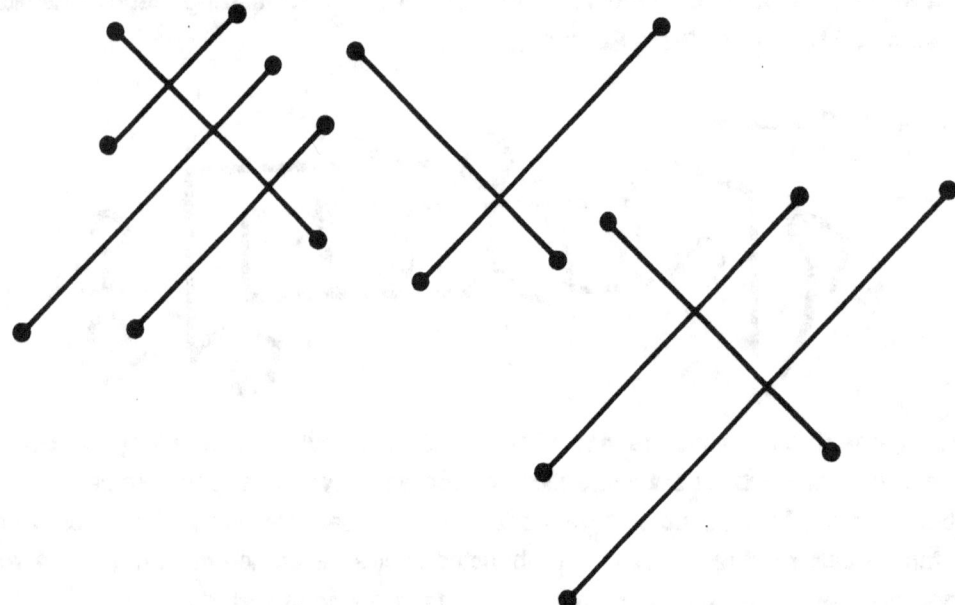

**21st Century Power: Bio-Tech.** - We are leaving the Era of expendable resources, like oil based products. The power of the future is renewable resources like Biologically Engineered products. The Dulce Genetic research was originally funded under the cloak of "Black Budget" secrecy, (Billions of dollars).

They were interested in intelligent disposable biology, (Humanoids), to do the dangerous Atomic, (Plutonium) propulsors and saucer experiments. They cloned their own little humanoids, via a process perfected in the Bio-Genetic Research center of the world, Los Alamos ! Now they have their disposable slave-race. Like the Greys (EBEs), the U.S. Government clandestinely impregnate human females, then removed the hybrid fetus's, also after about ninety (90) days, like the Aliens, (for more details about how the Government does their abductions, see volume two of this document, called the PULSAR PROJECT, ATR-25A, CODE: ARAMIS V - ADR 3 -24AT) then accelerate their growth in the lab Bio-Genetic, (DNA Manipulation) programming is instilled, they are implanted and controlled at a distance through regular RF (Radio Frequency) transmissions. Many humans are also being implanted with brain transceivers, these act as a telepathic channels and telemetric brain manipulation devices. The network, net was set-up by DARPA, (Defense Advanced Research Project Agency). Two of the procedures were R.H.I.C. (Radio Hypnotic Intracerebral Control) and E.D.O.M., (Electronic Dissolution of Memory). The brain transceiver is inserted into the head through the nose, they learned this from the Greys. These devices are used in the Soviet Union and the United States, as well as, Sweden. The Swedish Prime-Minister Palme gave the National Swedish Police Board the right in 1973, to insert brain transmitters into the heads of human beings covertly !

They also developed ELF and E.M. wave propagation equipment, rays which affect the nerves and can cause nausea, fatigue, irritability, even death. This is essentially the same as Richard Shaver's cavern "Telaug" mechanism. This research into biodynamic relationships within organisms, (Biological Plasma) has already produced a ray that can change the genetic structure, heal, or even kill. Shaver's cavern "Ben-Mech" could heal !

**Warning: Manipulation and Control** - Fear, Fraud, and Favor > > > > The Pentagon, the CIA, NSA DEA, FBI, NSC, etc. seek to capitalize on the beliefs of the American Public. The secret Government is getting ready to stage a contact landing with Aliens in the very near future. This way they can control the release of Alien related propaganda. We will be told of an inter-stellar conflict, but what looks real, may be fake. What is disinformation ?, is your attention being diverted by the strategy of a Shadow Plan ? I believe you already know the answer !

"Berkeley, Los Alamos Labs Chosen to Explore Makeup of Human Genome"

## APPENDIX A

## ACRONYM DEFINITIONS

| | | | |
|---|---|---|---|
| **AC** | ALIEN CRAFT | **NSA** | NATIONAL SECURITY AGENCY |
| **IAC** | IDENTIFIED ALIEN CRAFT | **NSC** | NATIONAL SECURITY COUNCIL |
| **INAC** | IDENTIFIED NUMBER OF ALIEN CRAFT | **DIA** | DEFENSE INTELLIGENCE AGENCY |
| **ACL** | ALIEN CRAFT LANDING | **CIA** | CENTRAL INTELLIGENCE AGENCY |
| **ACRL** | ALIEN CRASH LANDING | **FBI** | FED. BUREAU OF INVESTIGATION |
| **CRL** | CRASH LANDING | **UFO** | UNIDENTIFIED FLYING OBJECT |
| **ADR** | ALIEN DEFENSE RESEARCH | | |

**R&DB** THE RESEARCH & DEVELOPMENT BOARD   **ONR** THE OFFICE OF NAVAL RESEARCH

**AFRD** AIR FORCE RESEARCH & DEVELOPMENT

**CIA-OSI** CIA OFFICE OF SCIENTIFIC INTELLIGENCE

**NSA-OSI** NSA OFFICE OF SCIENTIFIC INTELLIGENCE

**MAP REFERENCES:** 65ATN-6

        MAP TRACT NUMBER 6
        LOCATION ON MAP (COORDINATES) 65 BY A.

(A-1)

# FORWARD FLIGHT

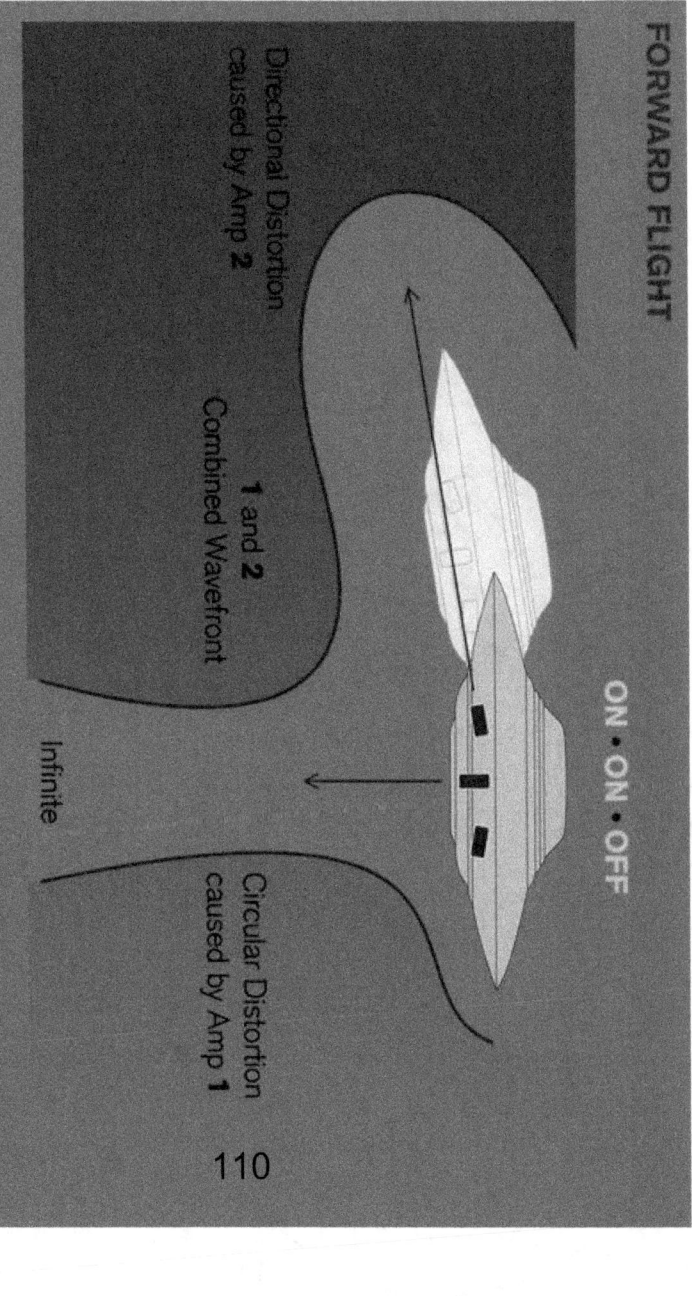

ON • ON • OFF

Directional Distortion caused by Amp **2**

**1** and **2** Combined Wavefront

Infinite

Circular Distortion caused by Amp **1**

In order to travel across the surface of a planet, such as in forward or diretional flight, the craft maintains altitude using one amplifier shooting towards the planet, while the other amplifier warps space-time in the desired direction of travel. The craft 'falls' into this depression, and moves forward in a way that the craft is always falling 'downhill' in the direction of travel. This dynamic accounts for the craft's wobbly appearance. In this mode of travel, only two gravity amplifiers are needed for flight.

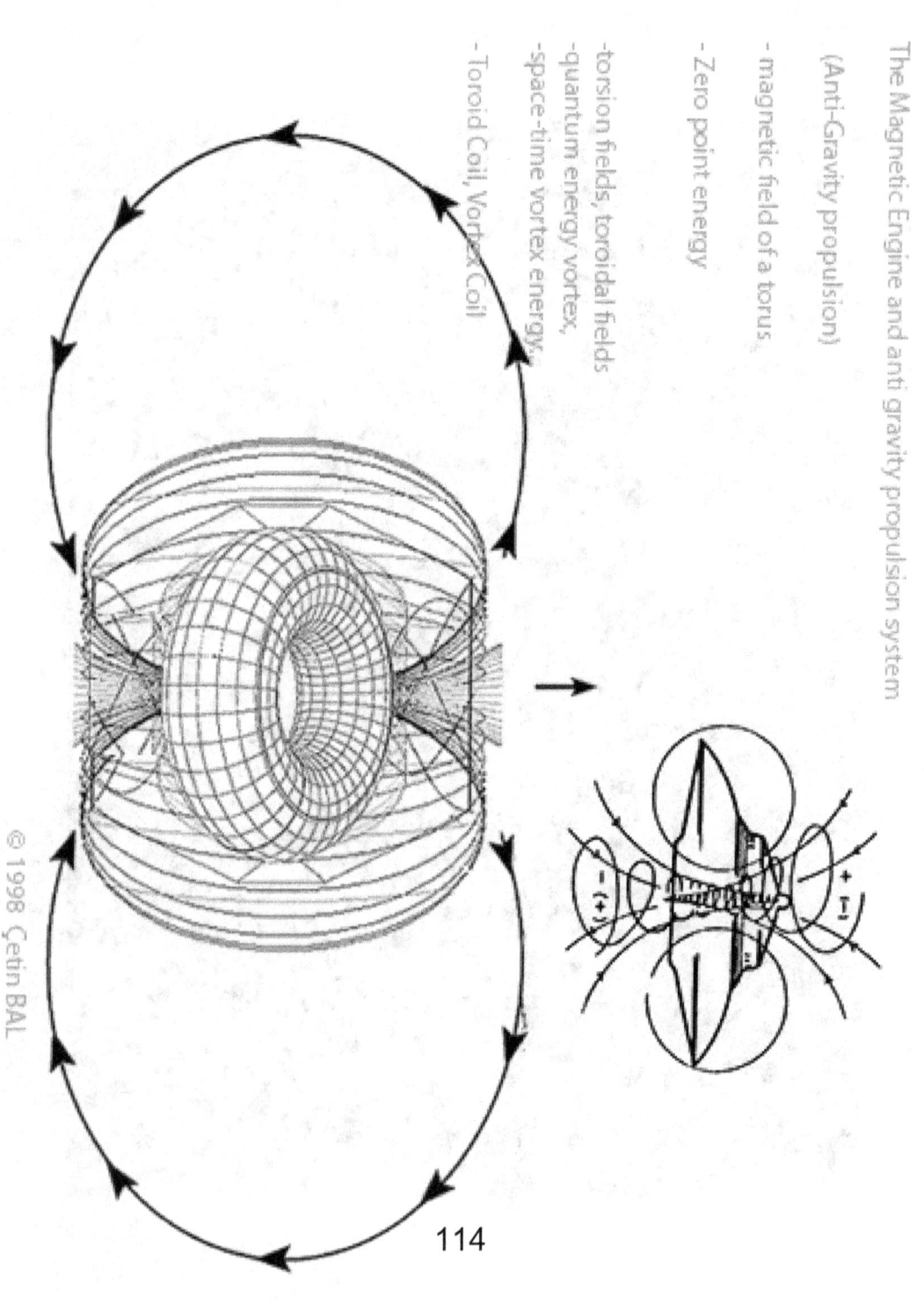

The Magnetic Engine and anti gravity propulsion system (Anti-Gravity propulsion)

- magnetic field of a torus,
- Zero point energy
- torsion fields, toroidal fields
- quantum energy vortex,
- space-time vortex energy,
- Toroid Coil, Vortex Coil

© 1998 Çetin BAL

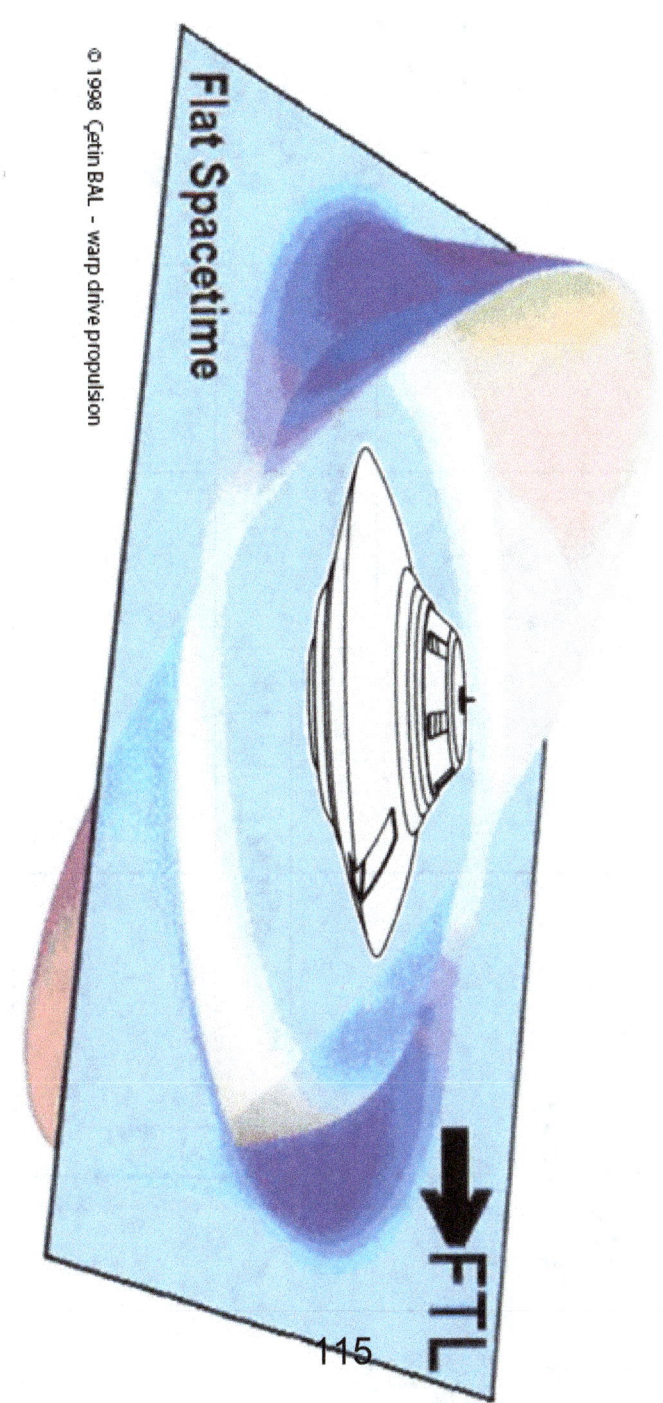

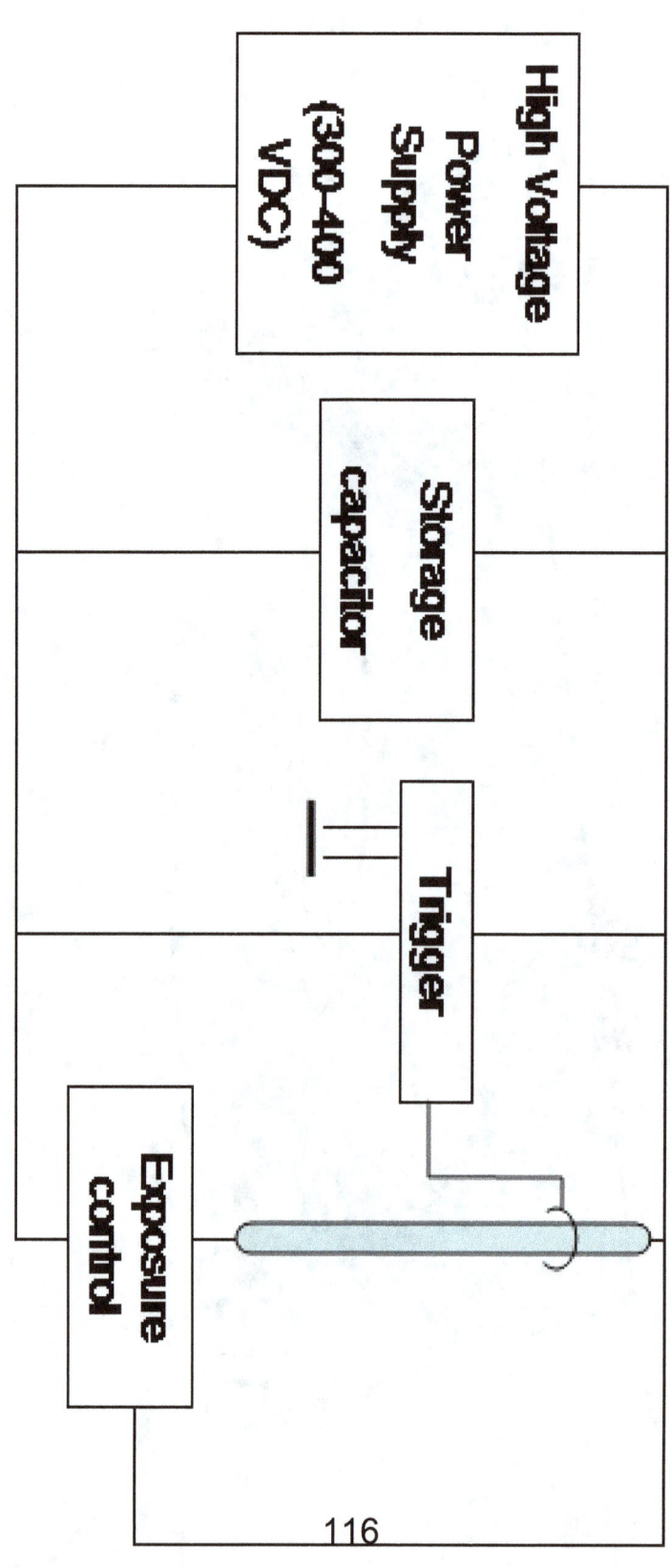

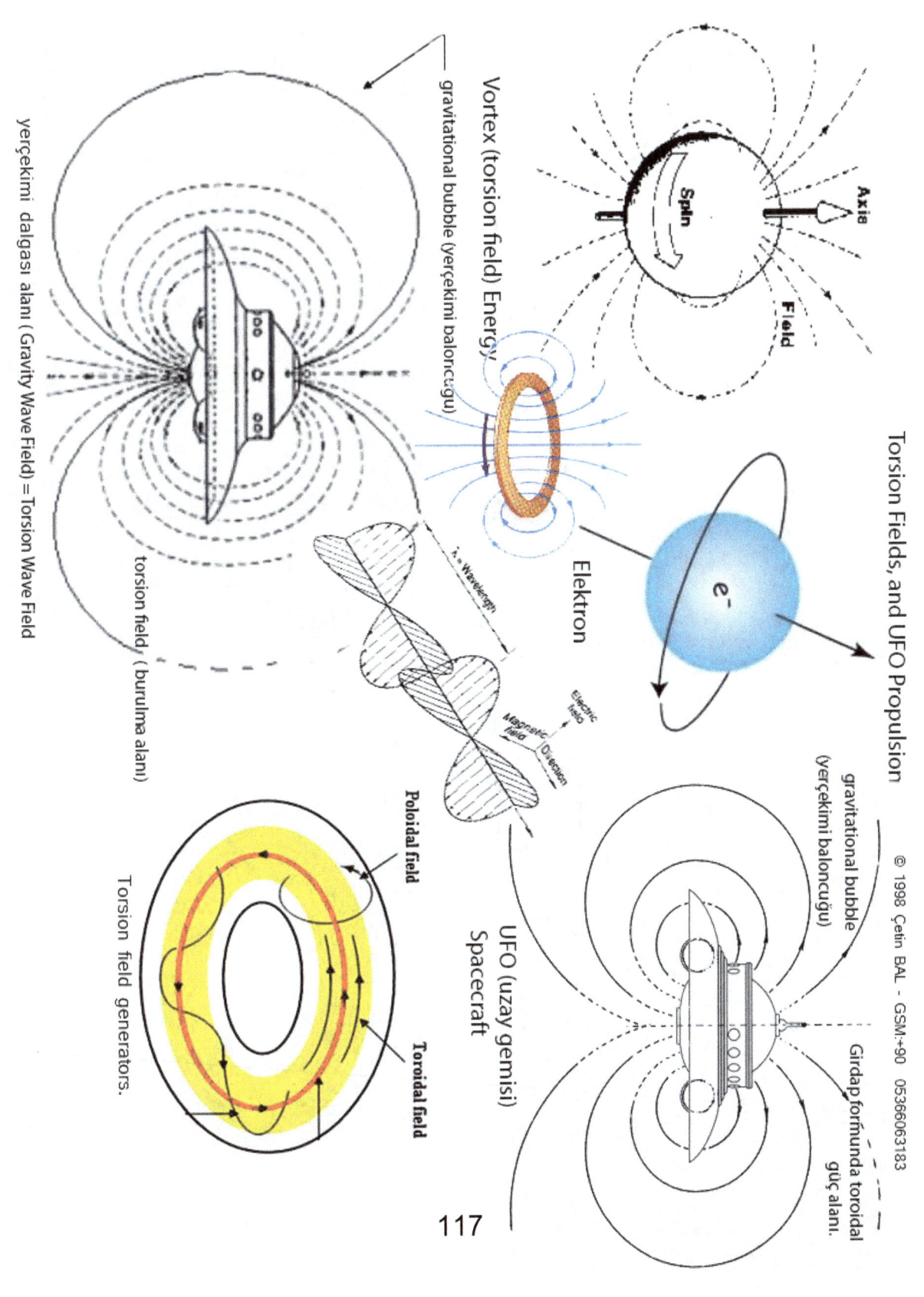

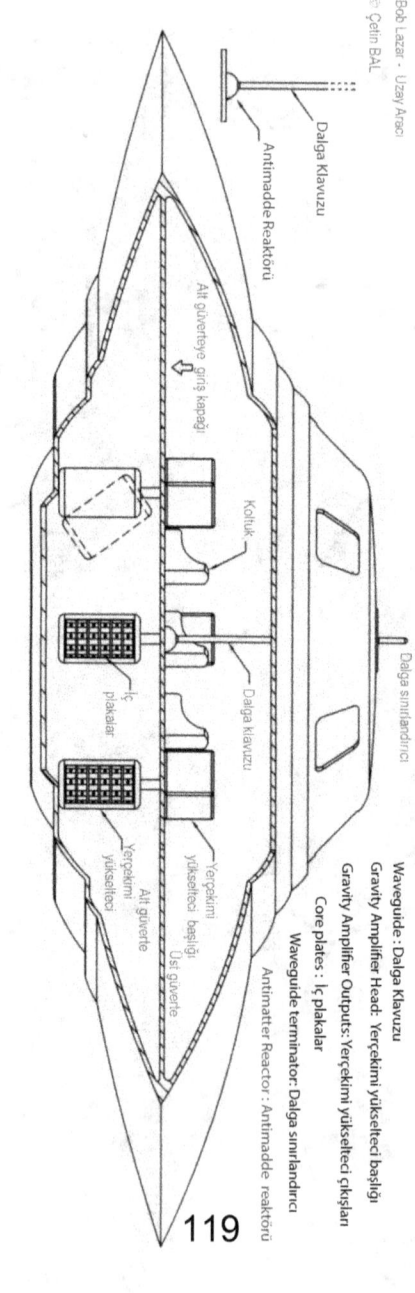

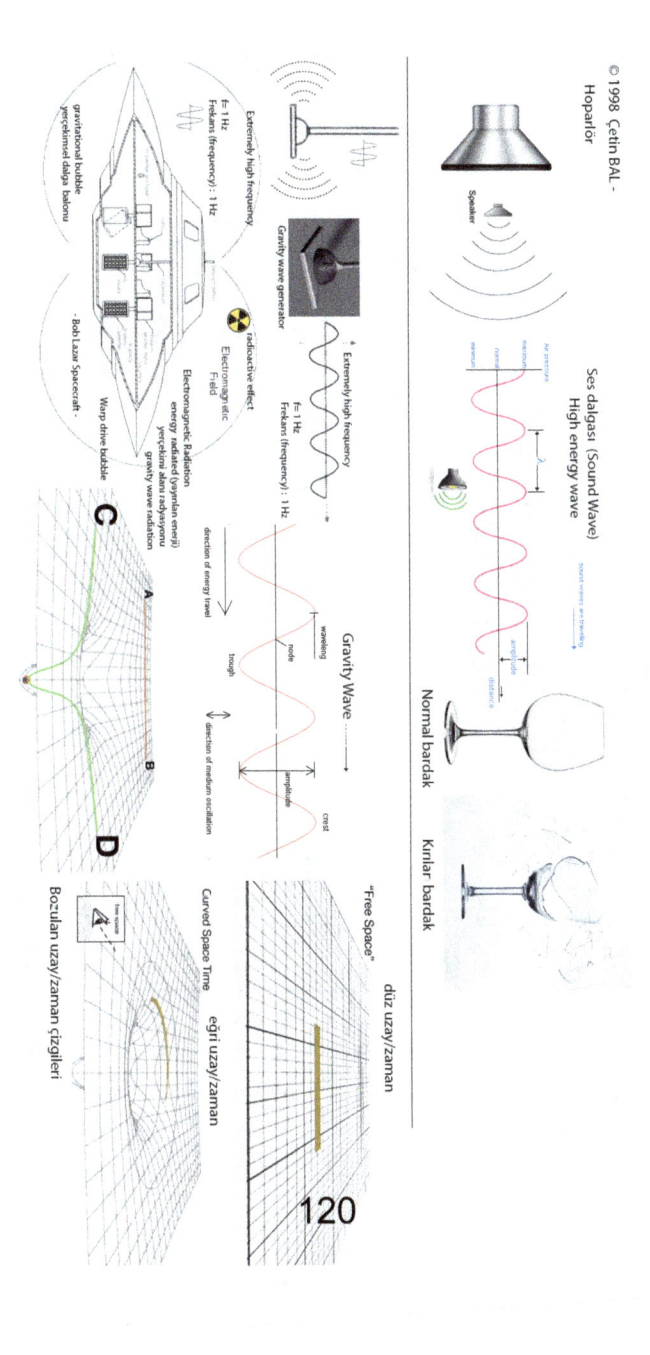

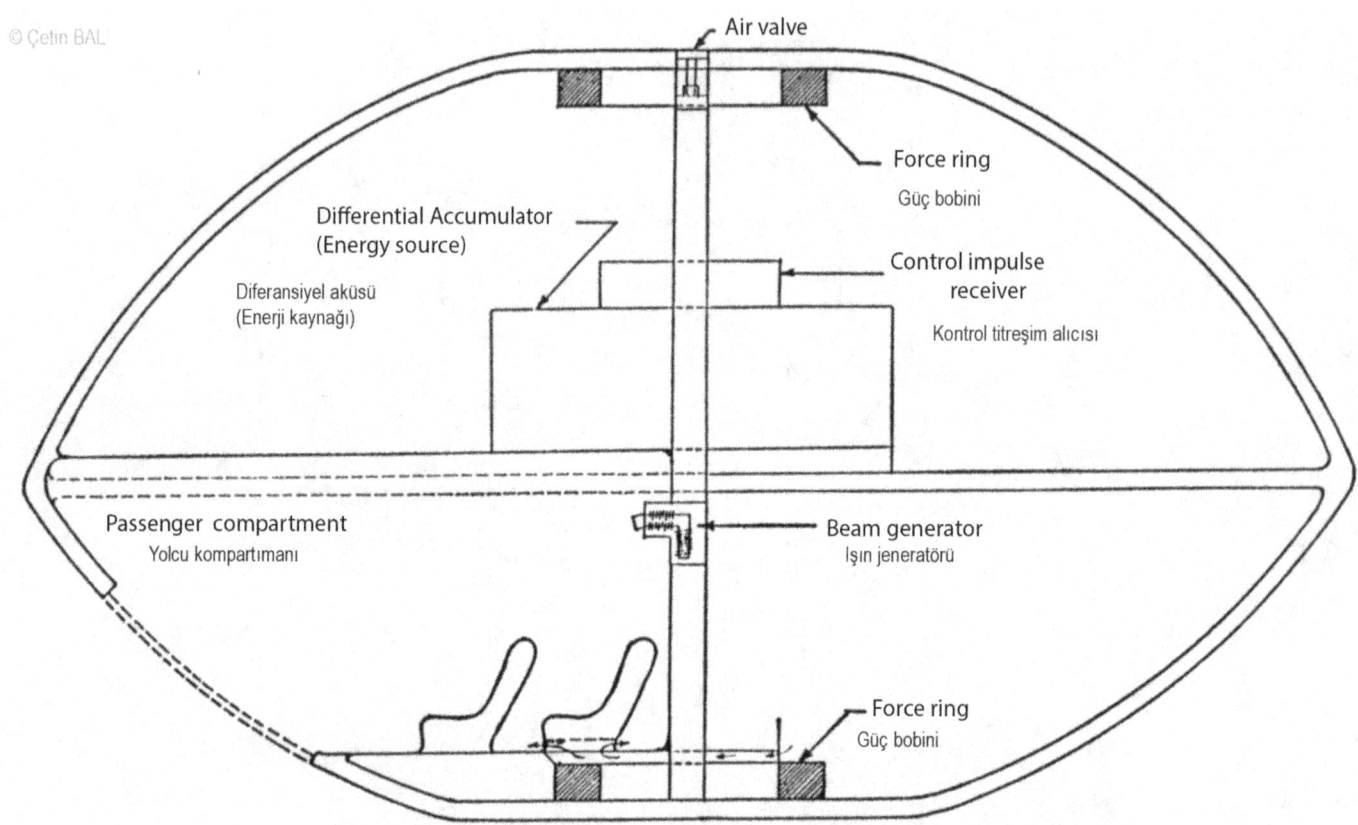

INSIDE OF SPACE SHIP DESCRIBED BY DANIEL FRY, PH.D.
Daniel Fry'ın bindiği UFO'nun kesiti - Uzay gemisinin içinin Daniel Fry tarafından tarifi.

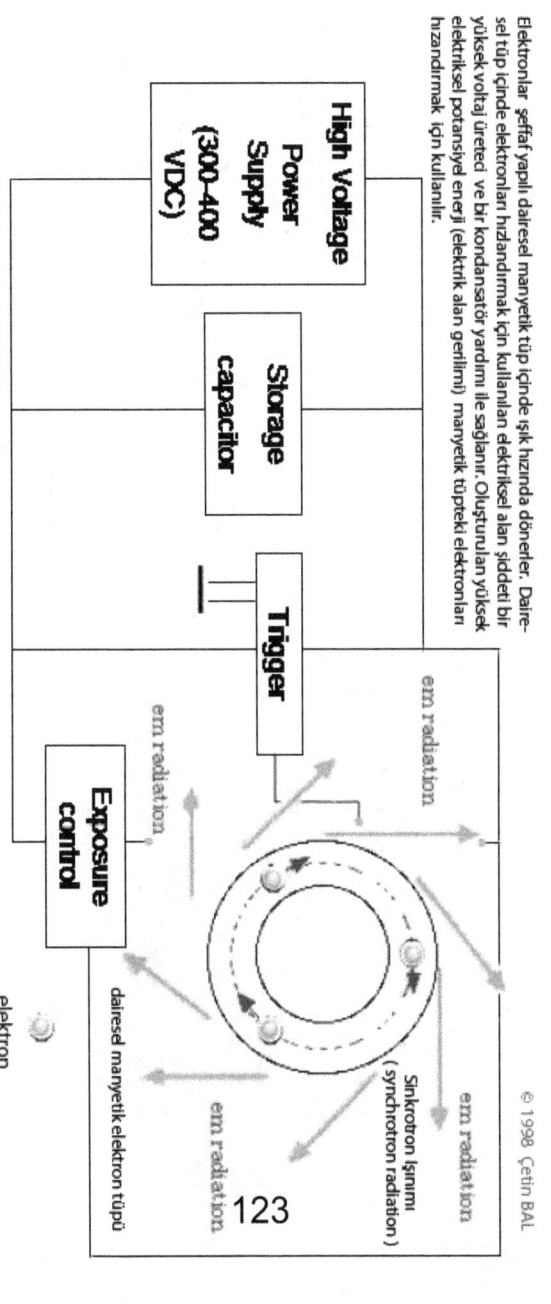

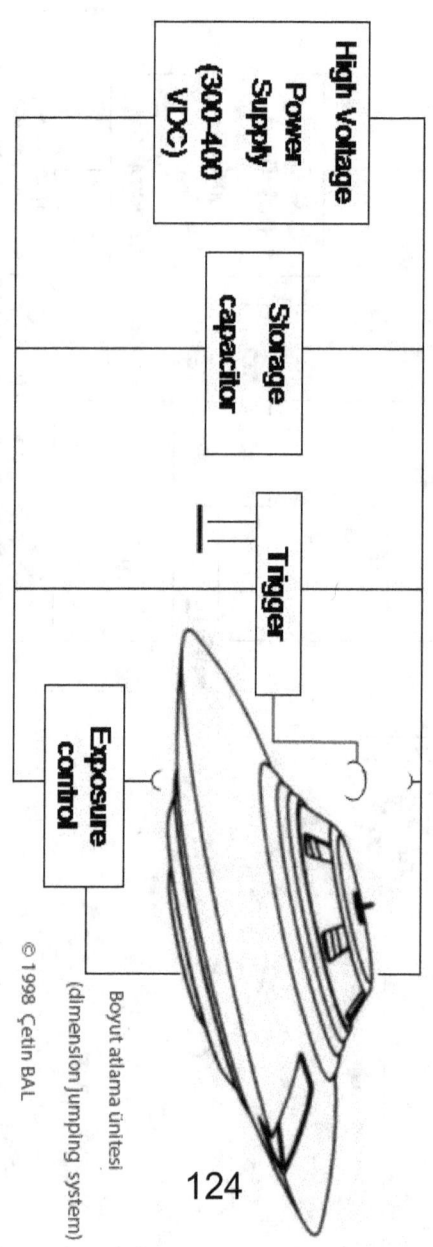

Boyut atlama ünitesi
(dimension jumping system)

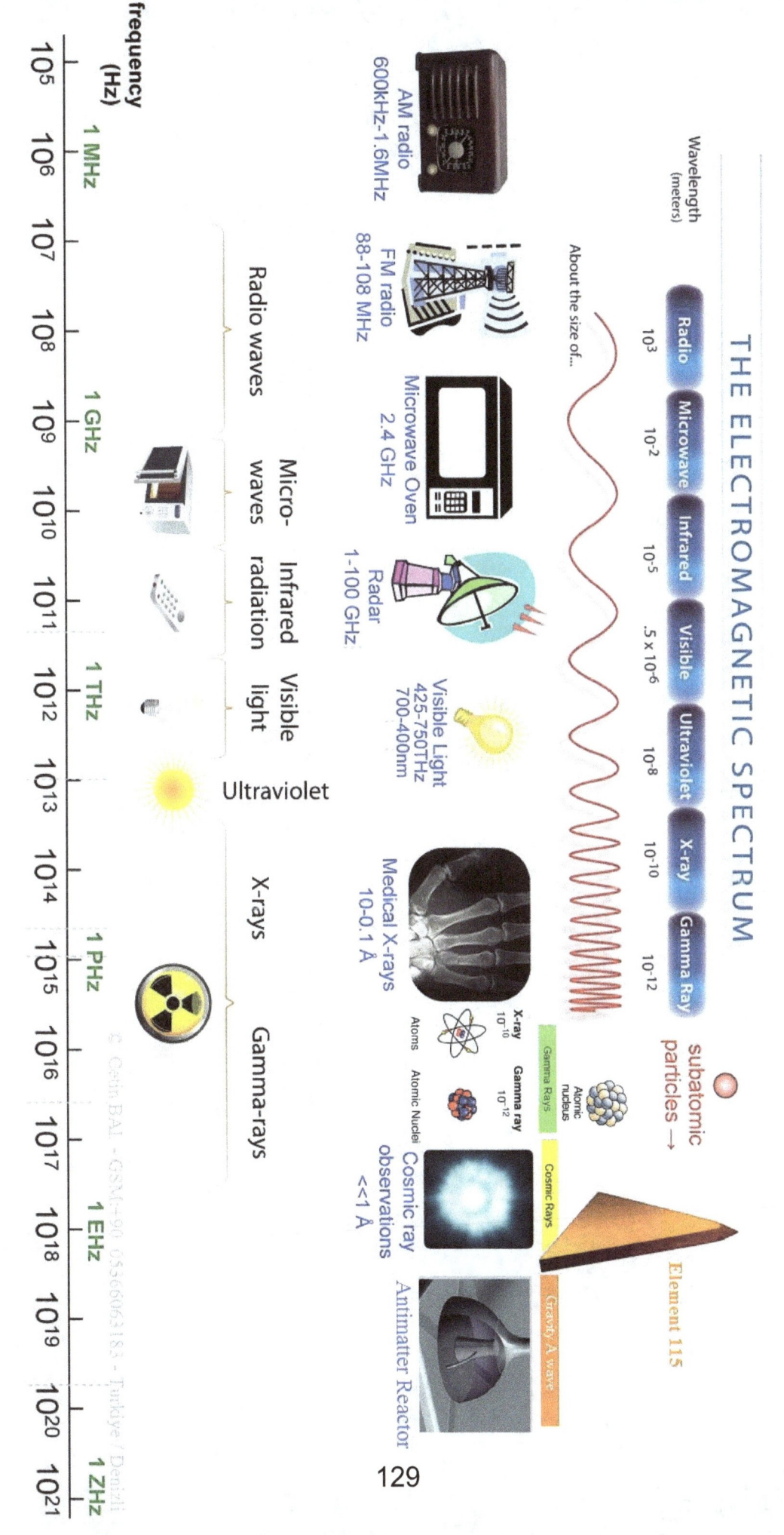

www.ingramcontent.com/pod-product-compliance
Lightning Source LLC
Chambersburg PA
CBHW080500220526
45465CB00006B/2330